IRISH TITAN, IRISH TOILERS

Revisiting New England: The New Regionalism

SERIES EDITORS

Siobhan Senier *University of New Hampshire*
Darren Ranco *Dartmouth College*
Adam Sweeting *Boston University*
David H. Watters *University of New Hampshire*

This series presents fresh discussions of the distinctiveness of New England culture. The editors seek manuscripts examining the history of New England regionalism; the way its culture came to represent American national culture; the interaction between that "official" New England culture and the people who lived in the region; and local, subregional, or even biographical subjects as microcosms that explicitly open up and consider larger issues. The series welcomes new theoretical and historical perspectives and is designed to cross disciplinary boundaries and appeal to a wide audience.

For a complete list of books that are available in this series, please visit www.upne.com and www.upne.com/series/RVNE.html

Joseph Banigan

IRISH TITAN, IRISH TOILERS

JOSEPH BANIGAN AND
NINETEENTH-CENTURY
NEW ENGLAND LABOR

SCOTT MOLLOY

UNIVERSITY OF NEW HAMPSHIRE PRESS
Durham, New Hampshire
Published by University Press of New England
Hanover and London

To the Rhode Island Irish Famine Memorial Committee

UNIVERSITY OF NEW HAMPSHIRE PRESS
Published by University Press of New England,
One Court Street, Lebanon, NH 03766
www.upne.com
© 2008 by Scott Molloy
Printed in the United States of America
5 4 3 2 1

FRONTISPIECE
An oil portrait of Joseph Banigan, ca. 1890. Courtesy of Phil Banigan.

Library of Congress Cataloging-in-Publication Data
Molloy, Scott.
Irish titan, Irish toilers : Joseph Banigan and nineteenth- century
New England labor / Scott Molloy.
    p.   cm. — (Revisiting New England : the new regionalism)
Includes bibliographical references and index.
ISBN 978–1–58465–690–6 (cloth : alk. paper) —
ISBN 978–1–58465–691–3 (pbk. : alk. paper)
1. Industrial relations—New England—History—19th century.   2. Banigan,
Joseph.   3. Woonsocket Rubber Company—Rhode Island—History—
19th century.   4. Rubber industry and trade—Rhode Island—History—
19th century.   5. United States Rubber Company—History—19th century.
6. Businessmen—United States—History—19th century.   7. Irish—United
States—History—19th century.   I. Title.   II. Title: Joseph Banigan and
nineteenth-century New England labor.
HD8083.N34M65 2008
338.7'6782092—dc22
[B]   2008010978

CONTENTS

HEEL PRESSING

PARTS OF SLIPPER

JUNIOR HEEL PIECE

STAY

LINING

PIPEING UPPER

Making Heels &c. of Boots

CUTTING UPPERS

SCIENTIFIC AMERICAN N.Y.

ACKNOWLEDGMENTS

I wish I could have said thank you to more people, but the material for this book was so disparate and sparse that I became a search committee of one. Joseph Banigan, the subject of this work, died suddenly at age fifty-nine in 1898. He had sold his rubber footwear empire in Rhode Island to the United States Rubber Company several years earlier. Although he became president of that monopoly, his subsequent estrangement and premature passing left his business records in the hands of a corporation anxious to obliterate his name and accomplishments. Despite searching high and low, I was able to find only a handful of primary documents.

Similarly, Banigan's widow died just a few years after him. Apparently the family papers were dispersed or destroyed, despite the fact that Banigan's four children from his first marriage all lived in Rhode Island. His second wife, Maria Conway, contested the provisions of her husband's will and may not have been on the best of terms with the stepchildren. Whatever the reason, only a few personal papers survived, even among his descendants.

Fortunately Banigan's legendary business career, from child laborer from Ireland to chief of a major American cartel, was the subject of numerous newspaper articles. But these stories, though they contain some revealing details, generally fail to explore other aspects of his life. I made use of an academic and collegial network to ferret out whatever else might be hidden away in scattered sources and tangential repositories. The seemingly barren landscape eventually yielded some fruit. And for the help I received in that quest, I am indebted.

On the home front, librarians at the University of Rhode Island filled my weekly orders for a potpourri of monographs from around the country. That library also yielded a mother lode of primary sources in special collections under the name of a contemporary of Banigan's, local rubber baron Samuel Colt. The Rhode Island Catholic Diocesan Archives contained some significant Banigan items. I am grateful for services provided by the Providence Public Library and the Providence

College Archives. Rick Stattler at the Rhode Island Historical Society Library also gleaned some wonderful ephemera relating to Irish immigrants.

Farther afield, I traveled to the Lynn, Massachusetts, Public Library, to MIT in Cambridge, and to the vast rooms of the New York Public Library, tracking down rubber industry journals. I visited the University of Massachusetts at Amherst to use the extensive microfilm records in labor and business history. Harvard University's Baker Library has a small but important collection of documents from the rubber industry as well as the regional reports of the original Dun credit reports. The venerable Boston Public Library owns a plethora of microfilm reels of local newspapers that helped in tracking the regional parameters of footwear enterprises. Thank you to all those anonymous librarians who assisted me.

Some invaluable material came from Rhode Island historians, genealogists, and collectors. The Reverend Robert Hayman, the masterful author of the history of the Catholic Church in the state, unselfishly provided a cornucopia of sources related to Irish life. The index to his two- (soon to grow to three-) volume religious work provided innumerable details that would not otherwise be easily documented. He generously shared material for a forthcoming study of Irish Americans in Rhode Island as well. His kindness was all the more remarkable for someone who runs a parish, teaches at Providence College, and still manages to research and write. Thank you, Father Bob.

Two "amateur" historians, who probably know as much about Rhode Island as any professionals, generously assisted me too. Tom Greene of North Providence amazingly provided the original bound volumes of the *Woonsocket Patriot* newspaper, saving my time and eyesight. He also supplied me with period maps of Providence. Our cohort, Russ DeSimone of Middletown, continually sent me copies of material relating to the history of the Irish in the state. Both enriched the project.

I returned to graduate school at Providence College in the 1980s while I was still driving a city bus for a living. There I studied with Dr. Patrick T. Conley, who opened up a world of local ethnic history. His masterful lectures, inspiring writings, and constant assistance sent me along a new life path, from just trying to change the world to writing about it too. His embrace of public history had a profound impact on my career. A pamphlet about the local Irish that he authored during the country's bicentennial, when he headed the state's commission (ri76) on that national anniversary, never leaves my side. I have multiple copies scattered around my home and office for easy consultation. Words cannot reflect his many contributions in so many areas. In the case of Joseph Banigan, Dr. Conley provided me with a copy of an undergraduate study about the rubber king written in 1979 by another one of his students, Berenice Blessing. That term paper, more of a Master's thesis, contained wonderful information that made my own quest much easier.

Two of my best friends during my undergraduate days at Rhode Island College, Dennis Cabral and Anne Duffy, later married and became enthusiastic about their respective Portuguese and Irish ancestries. They both visited Ireland frequently, and Anne became a talented Irish-American genealogist with roots in County Monaghan, Banigan's birthplace. We spent ten days in Ireland in the fall of 2005, including valuable time at the Public Records Office of Northern Ireland (PRONI) in Belfast. PRONI contained the records of the Shirley Estate where the Banigans lived until leaving in 1847 during the Potato Famine. I found rent roll records as well as the family's assisted emigration papers. I also learned a lot about life in that era and region by poring through innumerable other documents.

We spent several days in County Monaghan trying to track down the general area of the Banigan homestead, now pasture land. We visited the offices of the Clogher Historical Society and discovered its annual publication, filled with articles and information about the county's history. I could never have written this monograph without the prodigious research that so many authors contributed to that gem of a journal.

Just as important, we made many contacts with local historians and genealogists. Grace Moloney, the secretary who runs the Clogher group on a daily basis at St. Macartan's College, was a wonderful host. Thanks to Peadar Murnane and Theo McMahon, with whom I had corresponded before the trip. We also hooked up with Roison Lafferty from the Dublin area, who shared her considerable knowledge about Monaghan and an interesting Banigan artifact with me. Her husband gave us a bottle of good Irish whiskey!

Poor Dennis got to do all the driving on the left side of the road and, despite the research time, we managed to see other parts of Ireland, including my grandfather's birthplace in County Roscommon.

Noel and Marie Molloy (no relation) treated us like family during our brief visit there. Upon returning home, Anne continued to check facts and figures about the Banigans, educated me about naming practices in Ireland, and made connections to other researchers in the field. My seasoned companions made a research trip even richer than our historical discoveries.

Several other friends and acquaintances referred me to valuable articles and documents along the way: retired Providence police detective, George Pearson; Deb Elkins; Ann Harney from Virginia Beach, who alerted me to a very important, overlooked article; Patricia Raub; Al McAloon who knew all things Irish; Vince Arnold, the indefatigable head of the projected Newport Museum of Irish History; John Myers at the Providence City Hall Archives; Joe Hagan, the only person I know who is a Knight of St. Gregory, just like Joseph Banigan; Ann Galligan, who shared her work about St. Maria's Home in Providence; Derek Milewski, a graduate student at Cornell University at the time, who discovered a hidden Banigan enterprise; Jim de Boer, who wrote a wonderful undergraduate paper at Brown University about local discrimination against the Irish; Joe Sullivan and Joe Parys, two enthusiastic sleuths who researched the highly charged John Gordon murder trial in Providence in 1843 that discolored the perception of Irish Famine refugees in the minds of Rhode Island Yankees; and John Murphy, although a busy banker, took great interest in my research. Bob Whitcomb, an editor at the *Providence Journal,* kindly printed my many essays about Rhode Island Irish topics.

Several chapters contain material about the strikes against Joseph Banigan by the Knights of Labor in 1885. I presented the original findings at the Twenty Second annual North American Labor History Conference in Detroit in October 1999. Bruce Cohen, Jane Lancaster, Marc Stern, and Robert Weir read the paper and made solid suggestions for improving it. An expanded version appeared in the journal *Labor History,* 44:2 (2003). I would like to thank the editor at the time, Leon Fink, for his assistance and recommendations. Subsequently I decided to enlarge the piece into a book. Several friends and colleagues looked at individual chapters or the entire manuscript. My sincere appreciation to Dr. Paddy Duffy, the reigning expert on the geopolitics of County Monaghan at St. Patrick's College, Maynooth University,

Ireland; and Professor Paul O'Malley, who taught me Irish history at Providence College, for examining the transatlantic chapter. Special thanks go to several other exceptional individuals who digested the whole manuscript: Al Klyberg, retired director of the Rhode Island Historical Society, and Ellen Wicklum, my immediate editor at the University Press of New England. For service above and beyond the call of duty, no words can describe the professional, exhaustive critique by my colleague Professor Eve Sterne of URI's History Department. She somehow managed to dissect the monograph in great detail while bringing up her two babies and conducting a full academic life herself. I am forever grateful to all of you.

Although Joseph Banigan had four children, there are not many direct offspring living today. Phil and Marilyn Banigan, schoolteachers in Pawtucket, inherited several items, including beautiful, professional portraits of Joseph Banigan and his second wife, Maria Conway. John Banigan, Phil's father, filled me in before his death on the last years of the Banigan mortuary in St. Francis Cemetery, during several phone calls to Florida. Another descendant who took a great interest in his ancestor was Dr. James Brennan, an orthodontist in Cumberland. He shared an album of exquisite photographs of the Banigan mansion in Providence.

For the last twelve years I have also labored on another project: a monument to those Irish Famine survivors who eventually came to Rhode Island like the Banigans. The Memorial Committee, while not all historically oriented, have certainly evolved into experts. I never worked with a nicer group of people: Anne Burns, Ray and Sue McKenna, Father Dan Trainor, Cathy Miller, Rolland and Pauline Grant, Mike O'Rourke, Jeanne Chapman, T. J. Tullie, Tom Kelly, Mike Doran, Paul Cavanaugh, Bob Shure, Dr. Patrick Conley and Gail Cahalan Conley, Tom Gill, Bill and Nancy Gilbane Jr., Bob and Patty Hynes, former Governor Joe Garrahy, Clare Barrett, Jack O'Gara, Kathy Leonard, John and Maureen Andrade, Tom Trainor, and Ed Johnson. Along the way, while we were waiting to secure enough money and expertise to finish the granite and bronze project, four comrades in the group died: Sue Donnelly, Mike White, Dick Purdy, and Al McAloon. I am especially indebted to Dr. Donald Deignan, another classmate at Rhode Island College in the 1960s, who usually provided the dark context of the Irish Famine in tandem with my lectures on the refugees to Rhode Island. Don's contributions to the effort will never be fully known.

I also owe a great debt to the University of Rhode Island for providing me with a semester, and later a year-long, sabbatical; a URI Center for the Humanities Fellowship (financed by Shannon Chandley); and the opportunity to teach an Irish-American history Honors colloquium. The students in the course were exceptional, especially Sean McIntyre, who later attended school in Galway, Ireland. Along the way I received fellowships from the Goff Institute for Ingenuity and Enterprise Studies and the John Nicholas Brown Center at Brown University. Rick Scholl, the current director of the Schmidt Labor Research Center where I reside at the university, both encouraged my offbeat endeavors and performed yeoman work on the computer programming for this book. My thanks also go to the rest of my colleagues at the Center, its former directors Ted Schmidt and Terry Thomason, and the university's Provost, Beverly Swan.

Because this book took so long, as a result of my inability to easily find source material, I lost both parents during the period of its preparation. Ironically, my Dutch-Yankee mother was able to trace her own roots to County Roscommon, Ireland, much to the chagrin of my father, who thought he had a monopoly on that side of the world. My search also led me to my father's cousin Evelyn Horton, who, before she died, sent me a plethora of family information as well as the only picture I have of my Irish great-grandmother, Bridget Molloy. During the course of this work my daughter Kelsey decided to major in Italian at URI while my other daughter, Cady, helped me in all things related to the computer.

I have one other debt to acknowledge. I received a phone call a couple of years ago after I wrote a popularized story about Banigan in the *Providence Journal*. The message on the answering machine came in a strong, youthful voice. The gentleman said his father worked for "Joe" Banigan and he wanted to share some stories with me. I was immediately skeptical because Banigan died in 1898. I finally spoke with John Parker, my mystery caller, and asked him how old he was. He replied, ninety-seven! I drove to Warwick, Rhode Island, to meet this remarkable person and probe his encyclopedic memory of just about everything that his father, a plumber/steamfitter, told him about employment with Banigan in the 1890s. John asked if I was related to a trolley conductor, Henry Molloy, whom he remembered as a child from a place where he used to meet his father in downtown Providence. Henry was my Irish-born grandfather! John Parker died soon after giving me several interviews. He enriched the concluding chapter of this narrative, as well as my own life, with material that could only have come from someone close to the source.

IRISH TITAN, IRISH TOILERS

INTRODUCTION: SPOUTING

· ·

> *"I object to having our country made the dumping ground*
> *of the world's degradation. How shall we Americanize*
> *these people [Irish Catholics] before they foreignize [sic]*
> *us. How shall we save them before they sink us. How*
> *shall we Christianize them before they demoralize us."*
> —Dr. K. B. Tucker, *Philadelphia minister, addressing*
> *the annual Rhode Island Baptist convention, 1899*

> *"Show us the Irishman who ever invented an article or*
> *machine of merit, who has thereby been the direct cause*
> *of furnishing employment to thousands of people and*
> *incidentally doing something toward civilization."*
> —James Wilkinson, *Rhode Island inventor, 1901*

In March 1901 the scion of a colonial Yankee family in Rhode Island sent a blistering letter to the *Providence Journal,* underscoring the continued animosity toward Irish Catholics in the state. James E. Wilkinson, an inventor living in Providence, complained that many local newspaper articles and editorial missives, inspired by the misplaced carping of local Hibernians, attacked the English and their form of government. "Irishmen like to criticize the British—as well as Americans," he wrote, "because they cannot dictate to them; for Americans and Britons are their superiors in every way, and lead all others, and naturally those who come from boglands who know nothing, and think less, try to show they have brains by 'spouting.'" He continued with a challenge: "Show us the Irishman who ever invented an article or a machine of merit, who has thereby been the direct cause of furnishing employment to thousands of people, and incidentally doing something toward civilization."[1] He displayed the characteristics of British imperial thought throughout the long diatribe, almost bordering on a white-man's-burden attitude, usually reserved for English colonialism in Asia and Africa.

Unfortunately Wilkinson failed to demonstrate much understanding of recent developments in Rhode Island and the increasingly positive role played by Irish-Catholic immigrants and their offspring in local society. Nor did he display any historical knowledge of his own family's liberal pedigree: They donated land for St. Mary's Catholic Church in Pawtucket, Rhode Island, in 1828. At the time, David Wilkinson, an inventor, iron manufacturer, and important contributor to the state's industrial revolution, employed some Irish immigrants. The Protestant industrialist felt that a spiritual outlet for his workers would provide a socially conservative influence. James Wilkinson, David's heir and another inventor, seemed to be living off the prominent, sensitive contributions and reputation of his forebears, who had so actively advanced the takeoff of the factory system at Slater's Mill in the early nineteenth century.[2]

Although Irish-Catholic immigrants had settled in the state since the 1820s, and particularly after the 1840s Potato Famine, they usually arrived in such a marginalized condition as to preclude any rapid rise to prominence. Centuries of debilitating English rule and a sustained agricultural lifestyle translated into a threadbare existence in an urban and industrial setting without the requisite cultural and work skills. These immigrants also exhibited a hostile, chip-on-the-shoulder attitude toward the English, a stance that was more than mutual. The ruling Yankee majority erected, directly and indirectly, ethnic, political, and religious barriers that inhibited any immediate Hibernian success here. Historical animosity, as devastating to the human spirit in Rhode Island as the potato blight was in Ireland, decomposed further into vicious rhetoric, street battles, and even a local civil war in 1842. Wilkinson, the letter writer, correctly crowed about the considerable achievements and socioeconomic dominance of his own kind in the state in the eighteenth and nineteenth centuries. He must have peered in trepidation at the dawn of the twentieth century when Irish Catholics seized demographic control of Rhode Island, the first state in the nation with a Catholic majority, although not all of it was Irish. Wilkinson, in his bitter editorial denunciation, conveniently left out some notable accomplishments by local Hibernian inhabitants who faced overwhelming financial and psychological obstacles that Wilkinson and his ancestors had never had to surmount.[3]

Among Irish-Catholic immigrants, Thomas and William Gilbane

founded a construction company in 1873 that now operates on the international scene while still based in Providence. James Hanley presided over one of the region's most successful breweries in that same era, although one can understand why Wilkinson penned no accolades for that endeavor. Patrick J. McCarthy, a youngster who lost most of his family to disease after fleeing the Potato Famine, became the only immigrant mayor of Providence, an office he won just a few years after Wilkinson's poison-pen letter. Lesser lights who, at least in perspective, discovered commercial success servicing their "own kind" formed an emerging lower middle class in this period as well. Others learned trades and struck out on their own as independent craftsmen, started their own modest businesses, or joined labor unions to get ahead. Unfortunately, far too many just struck out, but they did not deserve Wilkinson's cheap, cutting invective.[4]

On the other hand, Wilkinson's most egregious omission was to ignore the industrial orbit of Joseph Banigan, the immigrant founder of the nation's most prestigious rubber footwear company in Woonsocket, Rhode Island, after the Civil War. Banigan was an accomplished inventor, entrepreneur, and employer. He became president of the United States Rubber Company—one of the country's major cartels—several years before Wilkinson spewed his venom. Banigan's family fled the Potato Famine during its worst manifestation in 1847—Black '47, in the folklore of Ireland. They arrived here in the worst of circumstances, and at a time of blistering discrimination toward Irish Catholics. Banigan attended elementary school for a year before becoming a full-time worker at age nine. For nearly half a century he rode a trajectory that trumped most other rags-to-riches stories in the United States. Some say he became the wealthiest man in Rhode Island; he was certainly the first Irish-Catholic millionaire there. Financial success is an easy way to define Banigan in a materialistic world, then and now, but his story is more complex in many other ways. If Wilkinson had dared include Banigan in his screed—and it would have been almost impossible in the nation's smallest state not to have known the story—Banigan's individual accomplishments would have contradicted the disparaging generalizations, the exception at least tempering the rule. Banigan accomplished everything that Wilkinson claimed the Irish failed to do: he held patents, employed thousands, and presided over the nation's first rubber cartel.

This book is a social biography rather than simply an account of a dynamic and charismatic figure. To a lesser degree, it is also the story of Banigan's "life and times" in Rhode Island, a once popular approach to biography that has fallen into disfavor in the history profession. Unfortunately, Banigan left no repository of business or personal papers. In some ways the chronological and psychological gaps lay an even greater responsibility on the writer to explain the temper of the times and understand how this figure from a dispossessed subculture fit into the local and national pattern of assimilation. This approach, although frustrating because of disparate and scarce primary material, allows us to reach into the recesses of history. The lack of information, especially concerning Banigan's formative years, forces the author to construct a parallel world of ordinary Irish immigrants in Rhode Island that puts at least a collective face on these people, even if it lacks the polish of a professional portrait. Furthermore, the story of what happened to this immigrant generation in general highlights the unexpected luster and texture of Banigan's journey. Although many historians continue to address the issue of people, places, and events that exist in the shadow of our national annals, the effort remains challenging. The evidence is often as thin and imperceptible as the thousands of invisible capillaries that fuel the heart of an individual organism, but just as important even if out of sight. We must try to investigate these elements regardless of the difficulty in finding traditional sources, or even of obtaining any primary material at all. On the other hand, researchers of the lives of most of the Founding Fathers, the wealthy, or the famous have a plethora of almost daily citations for these subjects, and they might complain that they are overwhelmed by the massive available records.[5]

In the mid-nineteenth century Edward E. Hale, a prominent Worcester, Massachusetts, clergyman who summered on the Rhode Island shore, wrote a series of eight sympathetic, sociological essays about Irish immigration that appeared in the *Boston Daily Advertiser*. Seeking to create an understanding of the Irish plight, he confessed that, despite innumerable conversations and interviews with humble arrivals, he could extract little information because he found so little substance to their lives in either country. He discovered that the Irish formed an impenetrable opinion that their homeland was above reproach, and would admit "nothing that should throw a stain upon it." Moreover, from all he could discern, everyday life in Ireland was so monotonous that it precluded anything out of the ordinary: "A year or two of the excitement of America," Hale concluded, "seem[ed] to sweep back . . . Irish life to the indistinctness of a dream." In an observation that got to the heart of historical sources, Hale identified an analytical problem: "Letters from Ireland are singularly unproductive. Of all the letters to emigrants which I have ever seen," he wrote, "I do not now remember one, which contained much more than congratulations that the reader had arrived in a land of liberty,—and acknowledgements of remittances, or requests for them. There is quite an animated correspondence kept up,—considerable in its amount, though from the ignorance of the parties, very small in proportion to the large numbers who emigrate." Hale inadvertently displayed his biases, but future historians would discover ways to extract meaningful ore from those bland missives. Still the minister's observations, even if not fully accurate, remain painful.[6]

The Wilkinson letter, on the other hand, devoid of Hale's sympathy a half century earlier, was more than just an attack on the Irish in Rhode Island and their way of life. The editorial reflected the author's own definition of success, one that harked back to a time when the colony of Rhode Island consisted almost entirely of British immigrants, marginalized, defeated Native-American peoples, and black slaves. Under the leadership of the brilliant and progressive Roger Williams, the colony's English population accomplished a mighty task by carving out a haven for religious mavericks of British stock from neighboring Puritan colonies, and enshrining freedom of worship and political independence in a remarkable charter in 1663. That document, produced on the heels of the brutal English civil war, provided a liberal political framework that sought to prevent religious conflict and foster greater toleration. Rhode Island's provocative experiment in this arena played a pioneering (and often unappreciated) role in the Revolutionary era, especially the genesis of the First Amendment. The state went on, with an equal flair, to lead the new nation's Industrial Revolution as well. When a second wave of immigrants reached Rhode Island's shores—the first major thrust from outside the confines of England, though still within the British Isles—toleration plummeted. Prejudices toward Irish newcomers came to reflect the same animosities that flourished in the Old World. Similarly, as Rhode Island and American society expanded, a system of class bias divided the ranks of the original Yankee colonists even from their own progeny, the economy separating groups into different castes and categories just as in England itself. Ironically, the fallout from that socioeconomic division energized skilled Yankee tradesmen who pioneered the labor movement in Rhode Island, causing as much annoyance as any Irish troublemakers in the nineteenth century.[7]

As Rhode Island's industrial elite emerged, it excluded most of the Yankee population, regardless of shared heritage and bloodlines; wealth came to trump ethnicity in this entrepreneurial society, particularly for immigrant newcomers. Manufacturers and financiers, often doubling as politicians, ruled the roost and measured achievement by their own yardstick. By 1840, several years before the Banigan family arrived, almost 60 percent of native-born white males in the state could not vote, because a land-owning requirement disenfranchised them. Understandably, immigrants had an even tougher road to respectability when so many native inhabitants could not exercise the most basic principles in a democracy. Under the leadership of a blueblood reformer in 1842 a shaky, informal alliance between a small number of Irish Catholics and a larger contingent of the Yankee poor ignited an unsuccessful uprising. That episode, explained in chapter 2, unfairly branded Irish newcomers just a few years before the Famine sent thousands to these shores. At the time the era of Jacksonian democracy, most states outside the South were liberalizing voting laws by eliminating the worst ballot restrictions, but not Rhode Island. Whatever accomplishments the Irish achieved locally always paled before the Yankee stranglehold in the arenas that counted: business, finance, office holding, and enterprise. Regardless of stellar performances by the Irish in such less-appreciated but valuable areas as politics, the Catholic Church, sports, entertainment, the military, literature, and the blue-collar labor movement, the local ruling class relegated Irish-Catholic success into second- and third-rate categories and even denied that some of these achievements deserved positive recognition at all.[8]

Banigan's story is not an antidote to the Wilkinson letter or the much larger sentiment against Irish Catholics that it represented. Banigan, along with the handful of his compatriots who successfully competed with the state's many notable Yankee entrepreneurs, held only a small place in the ranks of that triumphant crowd. In fact, in a modern academic study of Rhode Island's Gilded Age elite that identified 180 noteworthy individuals, Banigan was the solitary Irish Catholic, and one of only a handful of immigrants. Of course, these pioneering industrial Yankee peers had, in some cases, a two-century headstart in various political and business ventures before the main body of Irish Catholics even arrived. English immigrants and their offspring had molded the culture and the societal framework that determined the entrance requirements to upward mobility. The sheer number of Yankee fortunes overwhelmed Banigan, Hanley, the Gilbanes, and other respectable compatriots. In the Irish-Catholic community, of course, these few individuals served as models to be emulated. Yet the sting of discrimination lasted well into modern times, some would say until the 1960 election of John F. Kennedy as the first Irish-Catholic president of the nation, a man with strong Rhode Island connections.[9]

———————

Even as Rhode Island poised at the threshold of modernity, a Protestant theologian, Dr. K. B. Tucker of Philadelphia, spoke at the state's annual Baptist convention. In 1638/1639 Rhode Island had become the site of the first Baptist congregation in America, a symbol of the colony's liberal stance toward dissent and diversity. In 1899, just a few years before Wilkinson's invective, the minister sarcastically pointed out that there were twice as many Irish in the United States as in Ireland. "I object to having our country made the dumping ground of the world's degradation. How shall we Americanize these people before they foreignize [sic] us," he sermonized. "How shall we save them before they sink us. How shall we Christianize them before they demoralize us."[10]

The Reverend Bernard O'Reilly (not to be confused with the Providence-based first Irish-American bishop in New England of the same name), a native of County Donegal, provided a more compassionate description of his flock in a letter to the Woonsocket Patriot in 1868, about the same time Banigan moved to that city to begin his rubber footwear dominion. "My people are poor," the priest said; "there is not a single wealthy man among them. But they are rich in the exhaustless generosity of their Irish nature and in the traditional faith which deems no sacrifice too great when there is a question of their ancestral worship, or of the glory of God's house."[11]

The Yankees wrote the formula for success, but even if certain new-comers met those criteria, the invisible ink spelled out an ethnic-religious disclaimer. In order not to repeat the same biases inherent in the ruling paradigm, Joseph Banigan must be judged in a different way that does not provide him a free pass simply because he equaled or surpassed the attainment and prosperity of local Yankees. Banigan employed business methods as ruthless as those used by the most hard-bitten robber baron to undermine his competitors. To his credit, he hired his own kind almost exclusively in the Banigan mills, where Irish-Catholic workers performed miracles of production unexpected for a people so recently removed from the "boglands." The progressive Banigan and his union employees slugged it out in 1885 in a strike that rocked the Irish community and the Catholic Church, and exposed the fissures in this seemingly solid ethnic enclave, but also offered the promise of liberation for the mass of its members.

Exile from Ireland seemed less severe in good times when the boss was one of your own kind; but when the employer from Erin crossed the line and blurred the difference between sensitive emigrant and skinflint Yankee, the conflict intensified beyond the usual labor-relations divide. The old-guard corporate establishment erupted in glee at the tussle, skewering the Irish Knights of Labor and Banigan, the Irish king of capital, with a Shakespearian curse on both their houses. Yet neither Banigan nor his skilled rubber workers could escape the ethnic coil that bound both parties together for better or worse. Irish patrician and Irish plebian discovered that progress in Rhode Island and, by implication, the United States involved a religious symbiosis between titan and toilers in the larger hostile environment. Both sides needed each other.

In retrospect the trajectory began much earlier in Banigan's life but, with so little information, any explanation is speculative at best. Unlike many of his immigrant generation, Banigan could read and write. Yet the manuscript trail is spotty, as if the thousands of documents generated by his career had been devoured by a modern-day paper shredder. For years the Rhode Island Historical Society library maintained a "Banigan build-up file," a small collection of scraps of information, short sketches, and obituaries about one of the state's greatest success stories. Other historical repositories in the area saved very little about this remarkable entrepreneur. Only the Catholic Diocese of Providence holds a handful of original material.

Banigan may have been too young to remember much of his childhood in Ireland, although the oppressive history must have reverberated throughout the Irish Fifth Ward in Providence as it did in Irish-American communities everywhere. Retrospective articles mentioned, on occasion, Banigan's competitive zeal against Yankee business rivals, but there is no real analysis to explain the intensity. His achievements gained him permanent notoriety, but even this meager preservation rested on future success. He escaped the cohesive yet stifling environment of the ethnic enclave and seemed to evade the discrimination that crippled his compatriots by cavorting among Yankee inventors who worshipped the scientific method rather than a narrow class and ethnic ascendancy. Banigan flourished in this meritocracy. His later Irish-Catholic activism suggests a wider ethnic and religious resentment aimed at patronizing elites or the belittlement of his countrymen; but he almost never mentioned it publicly.

Banigan made his mark in the world of business and high finance but also played a crucial role in labor history during the Great Uprising in 1885 by the Knights of Labor at his plants in Woonsocket, Rhode Island, and Millville, Massachusetts. Among the scant source material, the entrepreneurial and religious successes remain enshrined, but the withering strikes are almost lost to history. No obituary, among the many death notices in 1898, mentioned the walkouts that rocked the local industrial landscape. One of the great dichotomies between the histories of labor and capital is the blackout that employees suffer when left out of the story of industrial endeavor. Business histories, more often than not, emphasize invention, mechanization, and entrepreneurialism. Labor activism, union achievement, and even ordinary, daily toil usually play an almost invisible role within the management framework. Labor history, on the other hand, cannot escape the gravitational pull by the business superstructure and historians unable to tell the story of workers without a business framework.[12]

The problem is magnified further when employees belong to a recently arrived ethnic group. A brilliant historian of Irish America in Worcester, Massachusetts, Timothy Meagher, concluded that blue-collar history also contains a subtle prejudice that forcefully separates the immigration question from the employment scenario. "Labor historians have been interested primarily in second-generation ethnics as members of the working class," he states, "when, how, and why American workers of varied ethnic background have, or have not, been able to act in concert on behalf of their own interests as workers, not when, how, and why ethnic communities evolve over time." The Banigan strikes in 1885 demonstrated the symbiotic knot of nationality and production. J. Matthew Gallman, another ethnic historian who eschewed the usual centrality of refugees, instead looks at the role of Irish emigration from Liverpool to Philadelphia, as a "story of worlds that they entered and the ways in which their presence helped to change those worlds." Certainly the exodus to Rhode Island created similar headaches for the already prejudiced Yankee ascendancy. Meagher also chronicles the slow but revolutionary changes in everyday Irish-American life measured over generations, and the impact those subtle modifications had on a host community. Less visible are the accelerated transitions in labor upheavals by the Knights of Labor in places like Wooonsocket and Millville, and the position of a successful role model like Banigan in the ethnic enclave but outside the usual boundaries of achievement. The Irish-American experience, as much as it is codified and explained, still contains many loose ends and surprises.[13]

THE IRISH BACKGROUND

*"They died in their mountain glens, they died along the
sea-coast, they died on the roads, and they died in the fields;
they wandered into the towns, and died in the streets; they
closed their cabin doors, and lay down upon their beds,
and died of actual starvation in their houses."*
—*W. Steuart Trench, land agent, Shirley Estate,
County Monaghan, 1840s*

*"The Almighty, indeed, sent the potato blight but the English
created the Famine."*
—*John Mitchel, Irish revolutionary, writing in 1860*

Joseph Banigan spent his childhood in a tumultuous County Monaghan, Ireland, in Ulster, from 1839 to 1847.[1] He was too young to have
experienced or understood much of the constricted world around
him, but influential events—including famine, metastatic poverty, and
the emigration flight itself—must have had an enduring impact on
the youngster and his family. Such experiences usually survived in the
memory of the participants and the folklore of local Irish-American
communities. Children born in Ireland or the United States heard
these legendary stories, embellished or not, that trumped the original
incidents themselves and entered into the mythological mainstream
of these ethnic colonies. As historian Kerby Miller intelligently surmised, all Irish emigration became involuntary and driven by English
vindictiveness within the belief structure of Ireland and its intentional
and unintentional outcasts.[2]

Banigan never seemed to say much about his personal or family history publicly or, if he did, it was not recorded; but through his donations he supported Irish patriotism and Roman Catholicism at home

and abroad throughout his life. He displayed a penchant for charities, secular and religious, that assisted children in orphanages, hospitals, day care, and abusive relationships—perhaps in reaction to his own haunted youth. Banigan's philanthropy and generosity earned him honors from Pope Leo XIII. Banigan visited Ireland many times and, on at least a couple of occasions, left liberal donations with clerics. Similarly, he contributed to most charitable Catholic ventures in Rhode Island. Irish patriots, who crisscrossed the United States on ubiquitous fund-raising and propaganda sorties, made obligatory visits to Rhode Island's large Irish population and personal visits to Banigan, who invariably wrote them a check and praised their efforts toward an independent Ireland, especially one of a parliamentary nature. The press often mentioned Banigan's roots but never went into much detail. Still, the Irish-Catholic framework and, especially, Banigan's life on the Shirley Estate in County Monaghan as one of the youngest members of a family caught in a debilitating system of tenancy, must have left its mark: the household of a small farmer slipping into the ranks of a cottier or laborer. To help us understand his childhood we must consider conditions in Ireland in general and County Monaghan in particular in the decade leading up to the time his family left for the United States in 1847 during the height of the "Great Hunger."[3]

Ireland

For several centuries the English and Irish engaged in vicious battles. Invariably, the Irish lost, as tribal warfare, invasions, conquests, religious conflicts, colonialism, famine, emigration, and revolution wrenched the island. Agonizingly, hopes for a unified Protestant and Catholic nation were dashed time and again. England parceled out seized Irish land to absentee landlords whose representatives subdivided the estates into several classes of renters. The tenants descended down the food chain from "strong farms" larger than fifteen to twenty acres, to middling ones consisting of ten to twenty acres, to "small farms" with lesser plots, to cottiers and day workers toiling as agricultural wage labor for others while tilling a potato patch on the side. For centuries the Irish rebelled, revolted, and mutinied against the Protestant "ascendancy" that controlled 85 percent of the land by 1700.[4] They practiced Machiavellian politics when the English were preoccupied in some other European or New World struggle, but to no avail. In response to these often clumsy machinations, the British, sometimes with Irish allies, imposed a series of draconian Penal Laws that peaked in the early eighteenth century and physically and culturally incarcerated the Irish in their own land. Approximately three-fourths of Ireland's population adhered to Roman Catholicism and suffered a second-class citizenship because of it.[5] The Anglican Church of Ireland, another imperial graft, hypocritically claimed to represent a large minority on the island.[6]

The Penal Laws sought to break the allegiance to Rome by the small but influential Catholic aristocracy, many of whom converted to the established Anglican Church as a way to hold onto their estates. Other impediments prevented the rest of the population from buying or inheriting land, receiving a public education, or holding political office under the laws' suffocating dictates.[7] Catholics faced major obstacles to joining the military, practicing law, engaging in commerce, voting, or attending mass. At one time or another other restrictions limited or prohibited the ownership of guns, horses, or other necessities of life in that era.[8] Much of Irish-Catholic society operated underground or in a parallel but illegal and hidden world. Joseph Banigan, born in 1839, was too young to understand any of this, although many of the legal barriers had fallen by that time. Scars continued to sear the Irish soul long afterward. Ironically, when Banigan's family emigrated to Rhode Island in 1847, they encountered a new kind of Penal Laws instituted by a Yankee ascendancy with another "Protestant Constitution." Similarly, hunger and disease reappeared periodically in poor, immigrant Irish-American communities in Rhode Island and across America in a kind of shadow Famine.[9]

The Potato Famine

The more traditional Anglo-Irish political battleground paled as the Famine made a dreaded early appearance in Ireland in the autumn of 1845, resulting in only a partial loss of the potato crop but providing a frightening glimpse of things to come. The ongoing segmentation of the Irish countryside into ever-shrinking parcels of land and the increasing reliance on the staple potato crop to sustain the dominant rural population were a disaster in waiting, especially in the west of

Ireland and other areas like South Ulster. Still the potato managed to seduce the Irish peasant. One acre of land, enriched with the abundant manure of livestock, yielded several tons of the vegetable. The potato grew unrestrained despite inhospitable soil in some places. The lowly spud supplied a nutritious if monotonous diet that included almost no fat but lots of energy. Once planted the potato required minimal attention until the fall harvest. When the Famine lingered until 1851, the English reiterated earlier complaints that the crop was a "lazy" one that encouraged the indolence of Irish farmers, a classic colonial assessment. Member of Parliament Evelyn Shirley, the proprietor of the largest estate in County Monaghan and landlord to the Banigan family, stated that "laziness is at the bottom of everything in Ireland."[10] Still the potato fed large Irish families; adult males allegedly wolfed down an incredible fourteen pounds a day. The pigs got a share as well. Amazingly, Ireland continued to export some cash crops of livestock and grains to England and the Continent during the Famine years, although the debate still rages over the actual amount of produce. There seems little justification or rational ideology to support the disbursement of foodstuffs produced in a debilitated country in need of those very comestibles to physically survive.[11]

The invasion by the blight was noiseless and invisible, the disease probably contained in manure imports to Europe from the United States. A damp and humid summer in 1845 created the perfect greenhouse conditions in Ireland for the poison. The wind carried minute fungal spores that landed on the lush green leaves of the potato plant; then a drop of moisture, soft rain, or dew detonated a biological catastrophe. The microscopic filaments of the deadly decay pierced the defenseless organism in an optimal climate for the encroaching microbes during Ireland's temperate summer growing season. The infection spread rapidly throughout the plant. Other spores moved independently underground to attack and putrefy the potatoes hiding beneath the soil. With the decay feeding the fungus, the deadly spores multiplied ferociously, contaminating nearby plants and fields and riding innocent breezes or raindrops across farms and counties. Sometimes the infection lay dormant in partially diseased spuds for a season, and farmers inadvertently increased their misfortunes by using the internally soiled potatoes as tubers to grow a new crop from the last harvest. The shoots that then burst through the soil sent a new generation of toxin into the air to recontaminate and expand the agricultural pathology. Even healthy potatoes, protected by a thick layer of soil or a lack of penetrating rain, contracted the blight after harvesting in a partially infected field. From 1845 to 1851 the pestilence seemed to enter the genetic code of Ireland's husbandry. Crop failures, followed by a partial seeding, sustained the vicious cycle. This agrarian epidemic destroyed the staple of the Irish food supply, the friendly potato, and decimated a population deprived of its major sustenance. Although the blight struck other parts of Europe in 1845, most countries farmed a wider variety of crops and suffered less overall damage than that inflicted on an entire class of small farmers and cottiers in Ireland who depended more on a single staple.[12]

In a population of over eight million, ironically swelled in the years before the Famine by the sustenance and ease of the potato crop itself, the blight practically attacked the people directly. Seven years of the Famine, alternating between partial and total devastation, tortured the Irish through starvation and associated diseases that claimed approximately one million victims. Hygiene dissolved and parasitic lice spread typhus. Peasants suffered from dysentery when they ate raw and undercooked foodstuffs. Various fevers spread invisibly like the blight itself among poor people who huddled together and, by doing so, inadvertently spread the diseases. The loss of the potato, rich in vitamin C, engendered scurvy. The different afflictions targeted the young and the old. Further, the drastic loss of caloric intake made men, in particular, susceptible to fever and unable to perform physical chores. Health care workers perished as frequently as peasants. Medical ignorance permeated all sectors of society.[13]

Outright starvation, however, was the exception. Approximately another million Irish, some of whom perished along the way, faced forced eviction and emigration to the United States, other parts of the British Empire, or points unknown. In the case of evictions due to nonpayment of rent, estate authorities simply propelled the downtrodden onto exit roads as agents destroyed the vacated hovels. Some plantation managers carried out the more humane yet still wrenching practice of assisted emigration elsewhere. There had been earlier instances in Ireland, but more inhabitants left between 1845 and 1855 than in the preceding 250 years. By 1890, 40 percent of the Irish-born lived in another country, as the Baptist minister pointed out in his acerbic remarks in the *Providence Journal* in 1899. To this day Ireland has never recovered the entire loss of those two million people, deported or dead, despite the passage of more than a century and a half. Those who stayed behind and managed to survive underwent drastic lifestyle changes.[14]

The most controversial aspect of the Potato Famine centered on the British response. The most haunting and long-standing refrain came from the Irish revolutionary John Mitchel, who helped lead the disastrous "Young Ireland" insurrection during the catastrophe in 1848. In a famous jeremiad he declared that: "the Almighty, indeed, sent the potato blight but the English created the Famine."[15] Irish livestock and cash crops of grain continued across the sea to England throughout the crisis, although the number of swine tapered off as they, too, depended on the remnants of the potato harvest for nourishment. The Byzantine decisions of the imperial British government, rent with ideological fissures, vacillated from honest assistance to a Darwinian zeal to depopulate Ireland. Sincere and insincere attempts to assist the Irish people dotted the countryside with workhouses, useless and useful construction projects, and soup kitchens. Some ventures were counterproductive, luring field hands to more "lucrative" public construction projects that in turn unintentionally jeopardized the harvesting of any remaining crops. The commemoration of the 150th anniversary of the Famine, beginning in 1995, sparked an outpouring of reassessment and scholarly attention unlike anything previously attempted. Irish historians finally subjected British policy to a withering academic assault. With the purest of objectivity one can only conclude that the British Empire generally deserved much of the blame, as John Mitchel claimed.[16]

County Monaghan

Historically, English soldiers and farmers bypassed "the bleak and miserable" landscape in County Monaghan for more fertile areas elsewhere. A demand for food and livestock jump-started the region's economy in the late eighteenth century during the Napoleonic Wars when England made an ambivalent move toward bringing commercial capitalism to Ireland.[17] At the same time, the county also developed a thriving flax and linen industry that augmented the more traditional agricultural staples. Home looms for weaving flax into linen nudged up the small farmer's income and, invariably, his rent. These tenants recruited cottiers as extra hands to increase productivity and, ironically, help pay the inflated bills occasioned by the once discretionary income from spinning. These permanent laborers only subdivided the land further. The population density in such small areas took a toll alongside the deterioration in the flax and linen industry, which petered out in the face of mechanized competition elsewhere in urban parts of Ulster. Monaghan became a "distressed county" even before the 1840s Famine. Food shortages appeared in 1800–1801 and later in 1817 and 1821. Crop failures and disease killed thousands. Furthermore, sporadic fighting between clandestine Protestant and Catholic groups, the genesis of which reached back generations, infected the area as much as the agricultural blights.[18]

The political movement known as Catholic Emancipation led by Daniel O'Connell flourished in Ireland, and County Monaghan too, triggering violence in the 1820s. Priests and parishioners joined the civil rights effort and targeted the local Member of Parliament, who was part of the Protestant ascendancy, titular head of the county's militia, and an absentee landlord. Monaghan became a vortex of activity as the 1826 election took on national implications. Despite the lack of a secret ballot (a perrennial issue in Rhode Island, including this era) and subsequent evictions, Catholic voters dislodged the incumbent. Evelyn Shirley, a more liberal-minded absentee landlord with some Catholic sympathies, won the seat after helping to register his own tenants to vote. The activity in Monaghan contributed to the legislation that undergirded Catholic Emancipation. The county continued to reflect the ongoing agitation in the rest of Ireland over tithes to the Anglican Church of Ireland, Father Theobold Mathew's temperance organization, and efforts to repeal the 1801 Act of Union between Ireland and England.[19]

On 7 June 1839, amid all this turmoil, Michael Joseph was born to Bernard and Alice "Bannikan," allegedly their seventh boy. Journalists sometimes used the apocryphal legend of the seventh son to explain Banigan's later fame and fortune in the United States. Along the way, the last name would change to Banigan and the youngster would use his middle name as his first. The family lived on the 40 square-mile Shirley estate in the civil parish of Magheross in the tiny "townland" of

Lisirrill, one of 56 civil divisions on the Shirley property in Monaghan. The 450 square-mile county (as compared to Rhode Island's 1,214 square miles of land and coastal waters) juts into present-day Northern Ireland like an arrowhead, but it has always been a political part of Ulster. However, the country's overwhelming Catholic majority assured its place in the Republic during partition in 1922. Banigan's father, Bernard, was one of 30 breadwinners jammed into Lisirrill, the smallest settlement on the estate, consisting of 274 acres, 2 roods, and 35 perches in the parlance of ancient English measurement. On average, each family shared less than 10 acres. Bernard Banigan also paid the smallest biannual rent of any of his neighbors, a minuscule 3 shillings, one of only two tenants paying less than 1 English pound in 1843. His rent would remain fairly constant until the family's emigration in 1847. Only four tenants paid more than 3 pounds. Similarly, the controversial bog fee for the Banigans was one of the lowest at 4 shillings, but it was slightly more than the rental charge for the land. On an estate map the few acres look well situated, nestled against a small body of water, Lough Namachree, one of the 184 lakes on the property. The paltry rent suggests that the ground probably contained swamp land, as did significant portions of the estate. The proprietor, Evelyn Shirley, petitioned the authorities in Dublin on several occasions during the Famine for special permission to use publicly paid labor to drain the many marshy areas that soaked his land. Bernard Banigan consistently paid his rental fees on time. He was listed in arrears, along with half his neighbors in Lisirrill, only in 1847, the year he brought his family to the United States during the worst effects of the potato blight.[20]

Joseph Banigan joined County Monaghan's population of approximately 200,000—about 370 per square mile, or 58 per 100 acres—almost all of whom lived on farms. (By contrast, Rhode Island, the first urbanized, industrial state in the United States, counted only 108,000 inhabitants on twice the amount of land at the time.) Professor Patrick Duffy, an expert on the geography of the region, called County Monaghan "among the most overcrowded countrysides in Europe" that teetered "on the edge of Malthusian disaster in the 1830s and '40s."[21] At the start of the Famine, he noted, "the resources of the land clearly had become too fragmented, farms had become minuscule, there were too many people with limited or no access to the land,

and fewer and fewer of these had access to non-farm incomes."[22] Several absentee English landlords controlled sizeable private estates. Evelyn Shirley, the relatively liberal proprietor of more than 25,000 acres—the largest plantation in the county—took up part-time residence a few years after his parliamentary victory in 1826. His agents helped divide the land into ever smaller plats in order to increase rents and pocket some of the expanded income, which averaged 21 percent above official land values. Ironically, the tenants themselves worsened the situation by strong-arming authorities to subdivide the land, sometimes against the better judgment of honest agents who understood the larger picture. Despite the significant rental income, approximately 1,000 estates fell into receivership in Ireland by the eve of the Famine. The number tripled by the end of the calamity in a kind of reverse Darwinism. To eliminate weak landlords and institute a forced merger of property under cover of the ongoing Famine, Parliament passed the Encumbered Estates Act in 1849. In a way, evictions stalked both ends of society.[23]

The duties of a landlord's agent, like the responsibilities of a police officer in any time or place, hardened management's operatives regardless of their prior sensibilities or outlook. Estate representatives collected rents, usually long overdue during the distressed early years of the 1840s, even before the Famine. They seized livestock, crops, and other property. They organized groups of constables, court officials, and even troops at times in order to fulfill their duties. They also dealt with petitions; some collective, others personal; to alleviate one problem or another under their jurisdiction. Once the Famine took hold, a crescendo of pleas asked simply for blankets.[24]

In 1843, four years after Joseph Banigan's birth, Shirley's unpopular agent, Alexander "Sandy" Mitchell, died of natural causes, an increasingly rare way to perish in the immediate years ahead in County Monaghan and Ireland. The local inhabitants celebrated his demise with bonfires and other festivities. Mitchell had introduced the unusual and despised practice of assessing annual fees for harvesting peat moss to use as cold weather fuel, a practice known as "turbary rights." His replacement, William Steuart Trench, described the atmosphere upon his own arrival: "the tenants on the estate were much excited, that they considered themselves ground down to the last point by the late agent; that they had for some time personally meditated an open rebellion against him, but now that he was dead, they determined to rise and demand a reduction of rent and the removal of the many grievances with which they stated that they were oppressed."[25] The aroused farmers, as well as those like the cottiers and laborers who owned a couple of acres or no property at all, wasted little time in educating Trench about these problems. The Shirley property accommodated about one person per acre, almost double the crowded conditions of the county itself.[26]

Trench came from England accompanying the landlord, Evelyn Shirley, the Member of Parliament, who now spent part of the year in Ireland. They stayed at Shirley's mansion close to the town of Carrickmacross. On 30 March 1843, a crowd gathered to demand justice, and Trench agreed to meet with the protestors in a few days. The tenants organized ten thousand inhabitants (probably with a few Banigans among them), about one-half of the estate's population, to confront Shirley and his new manager despite the landlord's attempt to indefinitely postpone the meeting. Shirley and Trench refused to lower rents, especially under the duress of such hostility, fearing that more militant demands might follow if they gave in to the initial clamor. When Shirley coaxed his new agent to somehow placate the "wild mob" the throng at first actually knelt in supplication but it soon turned angry, according to Trench, and dragged him to Shirley's manor over a mile away. "Those immediately around me," Trench later wrote, "mad with excitement, seemed only to thirst for my blood."[27]

Despite the seeming threat of death, Trench and his captors reached a stalemate: he pleading that he was just doing his job, they howling about injustices. But he promised to investigate complaints, claiming that he felt a bond with the peasants, who seemed embarrassed by their treatment of a total stranger. "I found myself unexpectedly possessed of a strange power over this wild and excitable people," he wrote.[28] Trench indeed seemed to have sympathy for their plight and the crowd finally released him. Some, however, discount the sudden and mystical peace accord and credit the local Catholic priest with pacifying the throng and saving the overseer.[29]

Trench, who published *Realities of Irish Life* in 1868, displayed a lively writing style and served as an important eyewitness for events in and around the Shirley estate during his tenure there, although he had his own obvious biases. Shirley, the landlord who had courageously allowed the construction of a Catholic school on his property the year Banigan was born and stood on a local platform there with temperance organizer Father Theobald Mathew, seemed to abandon his moderate-liberal mindset when his land payments ceased. The stuffy antiquarian history of County Monaghan that Shirley wrote had none of Trench's drama in it. Shirley described himself as a Saxon "who owed his Irish estate to the royal grant of Queen Elizabeth, and could not certainly boast of Milesian or Irish blood."[30] He sent bailiffs to collect rents, and at times he and Trench had employees commandeer the cattle of tenants in arrears and hold the livestock almost as ransom. Farmers learned to hide their animals, sometimes herding a few into their homes, which were practically stalls anyway. The authorities had no legal right to search inside the premises.

In 1843 Shirley eventually turned to a process known as "substitution of service" whereby a landlord could post a legal notice of arrears in a public place instead of actually serving the writ, an action that always presented a threat to the safety of officials. The tenants despised the new method, and their anger turned violent when the Protestant authorities tried to nail a summons onto the walls of the Catholic chapel in Magheracloone. To the parishioners from the Shirley estate, such a desecration sounded a call to battle. Tenants that inhabited the area turned out for hand-to-hand combat with the local police who accompanied the unfortunate bailiff placing the announcements. Confronted by an overwhelming but unarmed force, the constables shot into the crowd, wounding several and killing one. After the police retreated, the tenants turned the subsequent funeral into a protest demonstration. The Shirley estate then gave up any attempt to collect rents from the destitute and seething farmers, at least until the next harvest. During all of this, irate tenants dressed as Molly Maguires destroyed estate property during nighttime hours. These Irish vigilantes, operated under a host of names to mete out primitive justice.[31]

Only a year earlier, in 1842, a civil insurrection in Rhode Island, the Banigans' future home, pitted landholders who could vote against landless factory workers and a small number of pre-Famine Irish immigrants, an episode eerily similar to what was happening in Ireland with intense ethnic, socioeconomic, and religious overtones. In 1843 an Irish immigrant family allegedly killed a prominent Yankee industrialist in Rhode Island, further enflaming nationalistic and religious passions that led to a series of show trials and a hanging. Both episodes poisoned the atmosphere for subsequent Irish immigrants like the Banigans and at first seemingly made the upcoming move to Rhode Island a less than propitious one, as described in the next chapter.[32]

Trench claimed that after the battle of Magheracloone he personally visited dozens of tenants and listened intently to their grievances. He also learned that most of the farmers had pledged not to be the first to break local solidarity and pay the inflated rents. Trench talked some into voluntarily bringing in their cattle with the overdue payments. In a show of good faith he returned the livestock immediately. Whatever the case, Trench understood the long-term problem of the seemingly infinite subdivision of land on the Shirley estate and elsewhere into particles rather than parcels. When he presented a report on the dire situation to his employer, Shirley seemed offended by suggestions that he clear the land through assisted internal and external emigration. Perhaps Shirley feared the inevitable physical retaliation of such a revolutionary change or the loss of revenue if he evicted, however humanely, laborers, cottiers, and small farmers, all of whom contributed to his coffers directly and indirectly within this Irish geographical nightmare. When Shirley failed to respond to Trench's recommendations, his crestfallen subordinate soon left his employment. Although Trench appears to have portrayed himself fairly accurately in his semi-autobiographical volume, the introduction to the 1966 reprint stated that "his name is execrated in popular tradition. It was said that so evil was he that the rats invaded his grave and devoured his body."[33] The description of any agent's job in Irish history perhaps explains the dearth of saints among rent collectors. Trench would go on in life, and in his book, to recount the gruesome horror of the Famine in unimpeachable, eyewitness fashion. His contemporaries attacked his benign self-portrait as a fairy tale, but a private plan of action he submitted to Shirley to ameliorate the problems on the estate in a humane fashion has only recently come to light. To some degree it vindicates the maligned manager.[34]

The Famine attacked particular regions of Ireland so ruthlessly that areas less affected seemed, in comparison, to have escaped it altogether. Although some of the northern counties in Ulster avoided the worst effects of the calamity, the Famine took a toll there as well, irrespective of religion. Recent local studies of these territories have uncovered stress and disaster even in Belfast, with its industrial base and varied agriculture. In some quarters the interpretation of the political economy of the Great Famine outlined the "laziness" of the potato crop and its attendants as incompatible with the Protestant work ethic in Ulster. An ordinance survey in the mid-1830s ruefully remarked about County Monaghan that "there are no rotation crops: potatoes are the rotation crops."[35] According to Irish geographer Patrick Duffy, "Potatoes were widely cultivated in these areas where the smallholders and the cottiers were most numerous."[36] Seventy percent of the population in Monaghan relied fully or heavily on them. Local folklore fostered a belief that the Famine bypassed the north but smote the Catholic south in a sign of divine retribution. At times the English press hailed the Famine as a kind of modern-day "ethnic cleansing." Even some Catholics felt that the blight reflected holy discontent with Irish-Catholic mores.[37]

The Famine in Monaghan

The potato blight invaded Ulster and County Monaghan almost simultaneously with the assault on the rest of Ireland in the fall of 1845. Monaghan suffered in general as did the crowded Shirley estate in particular with its looming land and population problems dating from preceding generations. Most farm holdings on the manor measured a mere five acres or less, partially as a consequence of married children establishing homes on the edge of already shrinking properties. Almost 600 cottier families resided on Shirley's property, numbering more than 2,500 individuals, usually living in the poorest of domiciles with one room and no windows. They made up more than 15 percent of the estate's 18,000 inhabitants, who were almost entirely Catholic. The population had doubled since 1815, thanks in part to the nutritional wonders of the potato, and in spite of earlier agricultural problems. "Even in Ireland," Trench remarked in his private report to Shirley, "it has never fallen my lot to witness [land] distribution to the same degree and over such a large extent as I have seen on this property."[38]

Trench, who had moved to nearby Queen's County before the actual advent of the blight, later wrote that "I smelt the fearful stench, now so well known and recognized as the death-sign of each field of potatoes."[39] Local farmers let pigs forage among the rotting tubers.[40]

As the disease spread, the British government initially provided direct assistance, such as food depots for the sufferers; but a change in administration in 1846 promoted a laissez-faire approach that eventually emphasized public works. Some officials realized that whatever the cause of the Famine, the catastrophe paved the way, literally, for economic improvement and infrastructure development, from the interventionist Tory government of Prime Minister Robert Peel to the free trade regime of Lord John Russell. Even well-meaning efforts could not keep up with the changing nature of the Famine as each year passed and the blight jumped from one place to another, eventually reaching into far-flung and inaccessible rural areas. One local newspaper excoriated the inability or unwillingness of the authorities to act effectively: "Something must be done by the government before the winter and the spring of next year [1847]. They have time enough now to meet the exigency. There can be no excuse and the people must not be left to starve if the plague goes on."[41] Some local landlords in Monaghan, including Evelyn Shirley, reduced rents in exchange for labor to improve drainage on their properties, made donations to the poor, and even purchased "bad" potatoes from the tenants. In a letter to authorities in Dublin the subdirector of the Poor Union in Shirley's district remarked that Shirley "is very kind" to the unfortunate. Historians, on the other hand, challenge that interpretation, one calling the estate a "hellhole."[42]

Unfortunately, except for the soup kitchens that directly fed the ever-increasing destitute, workhouses seemed to spread disease, and public construction projects required physical strength that disappeared with the potato. Ireland became a farrago of relief committees organized by political zones. By the fall of 1846, the first harvest to fail entirely, Shirley sought funds from the government to help his constituents. The Irish Poor Law exempted from taxation those worth less than four pounds but even that figure allowed taxation of most small farmers with five to fifteen acres. Those least able to pay, who were already under tremendous financial pressure, faced spiraling assessments.[43] A petition to Dublin by local taxpayers in October 1846 also sought relief: "The patient endurance of our poor is equaled only by their own unexampled sufferings. These things cannot much longer endure and no authority or prevailing force can much longer stay the violence of a famishing population."[44] Inhabitants ate half-cooked potatoes that took longer to digest in order to slow gastric absorption and ease the pangs of hunger; others drank cows' blood, devoured rodents, and chewed seaweed when they were not barred from the beach by landlords. Trench again described the horror: "They died in their mountain glens, they died along the sea-coast, they died on the roads, and they died in the fields; they wandered into the towns, and died in the streets; they closed their cabin doors, and lay down upon their beds, and died of actual starvation in their houses."[45] Decomposing bodies littered the local landscape, as most people feared contamination if they neared the corpses to cover them. Domestic and wild animals displayed no such apprehension. In the decade between 1841 and 1851 the population of Monaghan decreased from 200,000 to 141,000 as a result of starvation, disease, and emigration: the worst losses in all of Ulster. The Shirley estate experienced a similar decline in the same period, from 18,000 inhabitants to 10,000.[46]

The Famine crisis loosened the few remaining social mores that protected tenants. When the major provisions of the English Poor Law of 1838 were extended to Ireland by Parliament in 1847, some 3,000 of the 18,000 tenants on the Shirley estate, including the Banigans, became eligible for cash-welfare paid by the landlord. One estimate claimed that 651 were already paupers in the local workhouse and qualified immediately for aid. Trench grasped the implications of the law, and Shirley probably understood that it was cheaper to help people leave than provide a stipend for them to stay, though he probably also feared jeopardizing the annual rent payments of more than twenty thousand pounds with a policy of emigration that might trigger an uprising by ejecting long-standing occupants. Financially "strong" Catholic farmers exploited the chaos to evict subtenants of their own. The Famine made it easier for landlords and renters to reach terms. The poor petitioners recognized that they still possessed some power to command assistance by threatening to continue to subdivide the land. An editorial in the *London Times* gave a more realistic analysis: "in a few years more, a Celtic Irishman will be as rare in Connemara as is the Red Indian on the shores of Manhattan."[47] Monaghan sent more than its share to foreign destinations. Between 1847 and 1852 Shirley paid about 1,300 tenants to leave and at least as many departed on their own. The emigrants surrendered their slivers of land, and Shirley, residing in his 27-bedroom estate, usually demolished whatever hovels they inhabited, allegedly destroying one hundred by 1849. A series of evictions that year also ignited some violence on the estate. The Banigans apparently surrendered their meager holdings to another neighbor for good. A quarter of all such small farms disappeared throughout Ireland during the Famine period. Like the Banigans, some who departed in 1847 did so as members of a family; others left on their own as young adults.[48]

The Banigans

Later biographical sketches of Joseph Banigan mention that the family lived in Ireland for three hundred years and left for Dundee, Scotland, when he was six years old, presumably in 1845, the first year of the blight. These accounts also declare that the Banigans stayed there for two years before returning to Ireland. One Irish historian concluded that "Scotland was a favoured destination for Monaghan emigrants and of the growing Irish population in the city of Dundee in 1851, no less than 50 per cent came from Monaghan and [County] Cavan."[49] Many laborers regularly traveled to parts of Britain on a seasonal basis to earn extra pay for harvesting crops. Dundee hosted a thriving textile industry, and mill owners often advertised in Ulster for weavers and spinners. Although records indicate that Banigan's father, Bernard, toiled as a "labourer," he or his wife may have had experience with flax, spinning, and linen, and may have sought employment in that industry since the area contained many "bleach greens." Some evidence exists that they had relatives in that area of Scotland as well. They may have left the land in Lisirrill to Bernard's brother-in-law and neighbor, Joseph Finnegan, who reunited with them later in Rhode Island.

At times the Shirley estate provided funds for emigration to Scotland as well as to overseas destinations. In the voluminous but partially indexed manuscripts housed in the Public Records Office of Northern Ireland in Belfast, there is also a petition from the "Banikan" family seeking expenses to the United States in 1847. Although the spelling is slightly different from even that of the rent roll, the names and approximate ages match up closely: the parents, Bernard, 38, and Alice, 36; the children, Catherine, 18, Patrick, 12, Michael (Joseph's first name), 10, and Mary, 1. The family received more than thirteen pounds, which included some funds for necessities on the trip, an off-setting mark against the stained reputation of MP Evelyn Shirley. Did the Banigans try to make a go of it in Dundee, only to return to Monaghan briefly to obtain assistance? Or did they run afoul of the English Vagrancy Act, aimed at resettling transient Irish back in their homeland? There is written evidence that the family petitioned Shirley to go to Australia as early as 1845, before going to Dundee. They may have tested the waters before deciding on the United States as a final destination, one that offered greater promise as well as a shorter and safer voyage. They apparently sailed from Liverpool, the usual port of embarkation, to New York or Boston, because Shirley had an enlightened policy of sending his tenants directly to their preferred destination, the United States, rather than on the cheaper trip to Canada. The Banigans eventually arrived in Providence, Rhode Island, in 1847, where at least one source mentioned they had relatives.[50]

Like the Banigans, those who escaped the ravages of the Famine and the quicksand of Lilliputian landholdings, faced a harrowing journey if they decided to settle in North America. Most voyages sailed directly to Canada, the shortest of routes, and since Canada was within the orbit of the British Empire, the cheapest vessels earned the reputation of "coffin ships" because of the dire health of their passengers and the on-board conditions that often rivaled circumstances back in Famine Ireland. About 16 percent of the emigrants perished during the trip, or died in quarantine or in a primitive hospital in Canada. Most of those who survived walked across the border to a final destination in the United States. Although there is no record of the Banigans' travel arrangements or on-board experience, they arrived in Providence and claimed a thin slice of economic salvation in exchange for an unexpected loaf of discrimination.[51]

Historians and many others debate the consequences of the "Great Hunger" and the dispersion of families like the Banigans with the same ferocity as the blight itself. Mary Daly captured the larger picture: "The Famine has, at one stage or another, been held responsible for almost every subsequent occurrence in Irish history."[52] The calamity certainly scattered thousands of refugees across the globe including Rhode Island, the smallest state in the United States, but larger than County Monaghan. The Banigans fled a tiny, rural homestead in a controlled agricultural setting in a polarized Ireland to a diminutive, politically regulated urban city in the United States. The Lilliputian boundaries and authoritative nature of each place were similar but the landscapes were as different as night and day. Growing up in a tumultuous Ireland seemed to energize Joseph Banigan later in his life when his philanthropic activities balanced childhood pain and grievances from a distant time and place. Adding injury to insult, the Protestant establishment in the United States always suspected that successful Irish Catholics had a strain of Yankee chromosomes in their ragged pedigree, just as American racists and some in the general public have often attributed white ancestry to successful light-skinned people of color. Banigan, the exception to the rule in the Famine exodus, undercut all the critics.[53]

RHODE ISLAND'S YANKEE ASCENDANCY

*"As you are now situated we think it would be more for your
comfort and happiness to leave Ireland and come live with
us, and as long as kind providence gives us a home you
shall be welcome to share it with us. . . . I think if they
[siblings] are industrious they can get a better living here
than in Ireland; for the youngest can earn as much as the
oldest can there."*
*—Irish immigrant John Fleming to his family in
County Kilkenney in a letter from Providence, 1836*

"OFFSET THE IRISHMEN AGAINST THE NIGGERS."
*—Law and Order Broadside, during the 1846 Rhode
Island gubernatorial election, claiming that Governor
Charles Jackson, a Whig moderate of Irish-Protestant
ancestry, would enfranchise Irish-Catholic immigrants*

Practitioners of many faiths called Rhode Island home almost from
the dawning of the colony. Roger Williams, the champion of religious
toleration who founded the refuge in 1636, welcomed Protestant dissenters—as well as Jews, Catholics, and even Muslims—to its shores, at
least in theory. A steady stream of Protestant settlers responded to the
welcome, especially radical sects like the Antinomians and Quakers
who often used Rhode Island as a base of operations to proselytize
among the hostile Puritans in neighboring Massachusetts. A small
number of Jews, fleeing religious persecution in the Iberian Peninsula
and South America, formed a thriving community in Newport in the
late 1700s. During the American Revolution, a century after Williams'
death, a temporary Catholic presence in the colony had more to do
with the arrival of French forces during the Revolution than with any

Irish-Catholic influx. All that would change after the first generation of Hibernian immigrants in the nineteenth century, without the earlier rhetoric of indulgence.[1]

Roger Williams initially fled the intolerance of neighboring Puritans in the Massachusetts Bay colony to escape a suffocating theocracy. He and a small band of followers sought "soul liberty," a way to freely pursue different religious and secular options. Ironically, the original settlers of New England, having thrown off the doctrinal shackles of the Anglican Church in Great Britain, harbored no sympathy for any form of worship besides their own once they arrived in the New World. Yet the history and nature of Puritan thought fostered a great degree of sectarianism and, at times, violent fragmentation. The articulate Williams, an ordained Anglican minister who had eagerly joined the dissident exodus to New England, soon questioned Puritan beliefs from several pulpits in Plymouth and the Bay Colony. He was tried and banished for his defiance. He avoided returning to England by establishing the tiny colony of Rhode Island in 1636 with help from the Wampanoag and Narragansett tribes. Williams treated the local Indians with respect, usually paying for any land purchases in fairer agreements than elsewhere in the colonies. He also studied tribal culture and language. Some high-powered Puritan dissidents sought refuge in Rhode Island as well, despite the immense theological differences between its various religious groups. These local exiles, by respecting one another's right to diverse religious viewpoints and practices, found a way of living in harmony that was highly unusual at the time. Williams scrupulously separated ecclesiastical ways from civil government.[2]

Massachusetts and Connecticut cast a covetous eye on the adjoining rebellious outpost that had no patent, charter, or other legal authority to exist. For a century these neighbors sought ways to incorporate Rhode Island into their own territory. Though meager in size, the colony contained valuable commercial waterfront that included an indented coastline much longer than its traditional boundary measurements. In the face of these antagonistic neighbors who were threatened by Rhode Island's contrarian religious and political independence, the colony embraced principles, like separation of church and state, that were more in line with the precepts of the coming American Revolution. In old England during the 1640s, Puritan extremism was igniting a brutal civil war that resulted in regicide of

Charles I, the widespread butchering of opposing forces, and Oliver Cromwell's savage invasion of Ireland. At the same time, in theocratic Massachusetts, religious intolerance and insecurity were leading to the imprisonment, torture, or execution of Quakers, "witches," and other dissenters in a bloody scenario that never materialized in liberal Rhode Island.[3]

Rhode Island remained a beacon of religious toleration and democratic rights for its inhabitants, a far cry from most political arrangements at the time almost anywhere in the world. During the mid-seventeenth-century upheavals in England, Roger Williams and Newport physician John Clarke lobbied Parliament for a charter. With the end of the Puritan Revolution in England and the subsequent reestablishment of the Stuart dynasty, King Charles II, son of the martyred Charles I, granted Rhode Island a colonial constitution in 1663 that enshrined its liberal tenets as a wedge against religious fanaticism. The colony would be a "lively experiment" that allowed freedom of conscience, widespread democracy, and religious liberty independent of any civil government. Most inhabitants could vote as freemen for representatives in the powerful General Assembly. In the colonial era, the landholding requirement for voting allowed about three-fourths of all white male residents to cast a ballot, reflecting the widespread ownership of land. Williams had wisely set aside plats for future inhabitants. The Charter of 1663 allowed four delegates to each of the original four towns, while subsequent civil subdivisions qualified for two representatives irrespective of population. One Rhode Island historian claimed the constitution, with its regular secular elections, legislative checks and balances, and amplified suffrage, created "a little republic."[4]

A century later Rhode Island played a Jekyll and Hyde role during the American War for Independence, being first in war (renunciation of allegiance to England on 4 May 1776) and last in peace (the final state to ratify the constitution on 29 May 1790). At times the colony roiled the British Empire by violating its trade and shipping regulations and its provisions for enforcing these. At other key points following the declaration of independence, Rhode Island's revolutionary fervor waned. After the Philadelphia Convention the new state refused to ratify the constitution, balking over the issue of import duties levied by the central government, threats to curtail Rhode Island's reliance on paper money, and fear of a remote government too far removed from local control. When a state convention finally ratified the constitution, Rhode Island's delegates instructed the congressional delegation to advocate establishment of a bill of rights that would provide more of the historic individual prerogatives its own inhabitants had enjoyed. Rhode Island finally became the last entity to embrace the constitution by the narrowest of tallies in 1790. The most diminutive state also decided, in its usual belligerent fashion, to retain the King Charles Charter of 1663 as its local legal blueprint. This ancient parchment, which had served the colony and its citizens so well for so long, stood until May 1843 as a legendary and still liberal icon long after most other states dumped their restrictive and conservative colonial charters and forged new documents replicating the national constitution at the local level, though keeping provincial concerns prominent. Rhode Island's unwillingness to replace or update its original charter would have grave and unintended consequences especially for the Famine Irish who eventually inundated the state in search of unskilled work and industrial jobs. Joseph Banigan, the young refugee from County Monaghan, would be one of only a handful to overcome those obstacles in his own lifetime.[5]

The Factory System

In the shadow of the American Revolution Rhode Island hosted another insurrectionary movement: the industrial revolution at Slater's Mill in Pawtucket, just north of Providence. For 150 years the state and the region had looked to the sea and the land for traditional ways of making a living. By the beginning of the nineteenth century, an economic transformation had dramatically undercut the use of both compass and plow. The revolutionary experiments of Samuel Slater, an English immigrant and skilled machinist, and Moses Brown, a cutting-edge investor, led, during the 1790s, to the foundation of the factory system in the United States. The ingenuity of Slater's Mill, with its fledgling industrial orbit in the surrounding Blackstone Valley of Rhode Island and Massachusetts, convinced many an investor to turn their entrepreneurial sights from oceanic horizons to the myriad inland tributaries that flowed into Narragansett Bay and the Atlantic Ocean. Fresh water streams and waterfalls, not saltwater and sea winds, would hereafter turn the wheels of the state's economy and alter the workplaces of the entire nation. Joseph Banigan would eventually set up shop in this valley of production six decades later, but first a political revolution spawned by the new order of toil would have to work itself out.[6]

The early factory system in Rhode Island, with only the English model to replicate, employed widespread child labor, including many siblings from the same family. This employment practice came to be known as the Rhode Island system, in a pejorative sense. Youngsters crawled beneath primitive machinery to retrieve bobbins and spools or rethread broken yarn. Any outburst by these young employees could result in the dismissal of the whole family. Despite corporal punishment and harsh working conditions, the owners paid in cash at a time when primitive bartering remained the norm in commercial dealings. Many farmers willingly sent their children into the factory at the dawn of industrialization to earn what husbandry could not: hard money. As mill conditions deteriorated over time, other states tried to avoid the excesses of child labor and punitive working conditions. By 1824 the country's first production plant in Rhode Island also hosted one of the first industrial strikes, led by teenage girls, weavers, who walked out following wage reductions and other cutbacks during a recessionary period of intense English competition. Although the "turn-out" ended in a compromise settlement, the young operatives managed to mobilize public support for their cause. The state's employment nexus now included a "them versus us" component that lasted as long as the factory system itself.[7]

The unintended political consequences of manufacturing that spread throughout the state over two generations, uncovered the brittleness of the progressive tenets of the King Charles Charter of 1663 once and for all. The General Assembly, which tied voting rights to the ownership of $134 worth of taxable land in 1798, maintained a broad-based electorate in a preindustrial period when most citizens still owned farms. Initially the workforce for the textile mills that spread along the area's water systems consisted of native-born children and young women of English descent. Demographic factors, the changing nature of work, and immigration patterns altered that. By the second generation of mill employment, ownership of property and real estate became a pipe dream for most operatives, who often had left the family farm and the right to vote behind in a rural birthplace. Skilled male artisans, while a minority in the factories, eventually found employment in ancillary businesses like tool and machine production located in mill villages. The new world order ripped the fragile, antique charter, which contained no direct amendment provisions but relied on Assembly delegates to keep up with the times through statute law. Time and again local reformers called for change via a constitutional convention to disentangle the right to vote from land ownership and provide fuller political apportionment for underrepresented urban towns and cities that now housed a growing number of factory hands.[8]

The Pre-Famine Irish

Patrick T. Conley, a pioneer of Irish-Catholic historiography in Rhode Island, wrote that "during the nearly sixty years from the arrival of the French in 1780 until the latter part of the 1830s, Catholicism was not only tolerated in Rhode Island, it was often cordially received."[9] Many other historians echo these sentiments of a kinder, gentler period in state and national history before the inundation of Famine refugees smothered most indulgence. Conley cites the traditional factors in the deteriorating position of Irish Catholics and the Yankee backlash in that later era: waves of immigration; increased poverty and decreased education among the newcomers; and the perceived threat of corrupt, urban political machines run off the backs and votes of Irish refugees as publicized in the experiences of New York City. Rhetorically, at least, the weapons of hatred had been loaded earlier than most historians appreciate. Furthermore, Rhode Island possessed the counterpart to an earlier century's witch hunters. Henry Bowen Anthony, a Whig official, became the editor of the increasingly influential and nativist newspaper, the *Providence Journal,* in 1838. For almost half a century, Anthony, from a number of political and editorial positions in the state, castigated Irish Catholics and other immigrants and incited violence against them. He organized opposition to civil rights for the Hibernian population by exaggerating their participation in the agitation leading to the Dorr War in 1842. Although he occasionally struck a conciliatory note, Anthony remained an unmitigated bigot.[10]

Socioeconomic changes in Ireland at the end of the Napoleonic Wars and the recession of the U.S. economy at the same time, following the War of 1812, dovetailed to negatively change the patterns of Irish immigration to the United States and Rhode Island. The shrinking quilt of land rental throughout Ireland as exemplified in Banigan's County Monaghan ignited a pre-Famine burst of immigration as some peasants perhaps sensed the forthcoming collapse of Irish society and escaped the impending crisis just as ship fares tumbled fortuitously. The few who journeyed here before the debilitating effects of the sustained potato famine unquestionably arrived with greater skills and fared better than their counterparts who came in a discombobulated state during and after the "Great Hunger." In 1828 when the first permanent parish was established as St. Mary's in Newport there were only 600 scattered parishioners in a state population nudging the 100,000 mark.[11]

John Fleming, a native of County Kilkenny, arrived in Newport in the 1830s. Upon the death of his father in Ireland in 1836, he and his wife invited his mother, a widow, to join them: "As you are now situated we think it would be more for your comfort and happiness to leave Ireland and come and live with us, and as long as kind providence gives us a home you shall be welcome to share it with us." He also advised her to bring his sisters and brothers too: "I think if they are industrious they can get a better living here than in Ireland; for the youngest," he predicted, "can earn as much here as the oldest can there."[12]

Fleming probably worked as a day laborer at Fort Adams in Newport, moved to Providence and operated a tavern in the Irish Fox Point section (Corky Hill), but appeared later in the city directory as a laborer. He seemed upbeat about life in Rhode Island: "We both enjoy very good health and like this country very much and we have the pleasure of hearing Mass every Sunday." Harsh reality soon changed this idyllic picture. Fleming arranged passage for his extended Irish family from Liverpool, England, in March 1837. (This usual port of embarkation for the United States was also used by the Banigan family exactly a decade later.) Fleming's mother and five siblings sold their meager belongings in Ireland to pay for their passage and buy some provisions for the voyage. Upon arrival at the British port town they discovered the ticket agent had no knowledge of them or of any prepaid tickets. Fleming had mortgaged the tavern and relied on a prominent Providence merchant to make the travel arrangements. The Protestant factor either bungled the transaction or swindled the would-be immigrant family. Fleming eventually sued, but most of his family returned to Ireland from Liverpool, brokenhearted and broke. His mother died on the journey. Fleming finally bankrolled the tickets again, but only two siblings took a second chance and finally arrived safely in Rhode Island. Whether the law suit ever progressed is unknown. At the time immigrants could not institute legal action in Rhode Island without a native-born sponsor. The agent, who owned a whorehouse where the chief prosecution witness in a notorious upcoming Irish immigrant murder trial lived, was also the brother of one of the state's Supreme Court justices.[13]

Pre-Famine arrivals like John Fleming usually found jobs on back-breaking public work projects: the construction of Fort Adams in Newport; coal mining in Portsmouth, or the erection of the columned Arcade in Providence, the nation's first "shopping center." Other endeavors included the laborious building of the forty-five-mile-long Blackstone Canal from Providence to Worcester, Massachusetts, and the laying of track beds for various railroads in the 1830s that would pass through Rhode Island and crisscross the New England region. Irish women found employment as seamstresses, domestic servants, and cooks. Men and women alike also toiled, as generations of succeeding immigrants would, in the sweated textile factories that consolidated the state's notoriety as the first urban, production-oriented entity in the nation. Initially most Irish Catholics faced discrimination when applying for factory employment, until conditions deteriorated in the mills and native-born farm children turned to other lines of work. The fabric industry spread to nearly every nook and cranny in the state that featured a stream or waterfall for energy. The subsequent and quick introduction of the power loom and steam engine in the 1830s provided greater geographic mobility to industry, although most mills stayed near a stream for a variety of purposes, especially waste disposal.[14]

In 1824 work began on Fort Adams, part of a federally sponsored national project to protect the U.S. coastline from possible British incursions like those experienced during the War of 1812. The facility, a massive granite and gravel project, would intermittently employ several hundred Irish laborers and masons until its completion in 1842. Some of the social intemperance that already stained the European view of the Irish in Yankee Rhode Island would manifest itself among

Bridget O'Farrell was the daughter of John O'Farrell from County Kilkenny, Ireland, who came to Rhode Island to work on the construction of Fort Adams in Newport around 1829. Bridget was born in December 1834 and is shown here on her wedding day in 1852 with her husband Patrick Guerin from County Limerick. The O'Farrell family left Newport about 1835 and moved to Iowa. Bridget and Patrick eventually settled in California and raised 11 children. She died there in 1888. The stunning tintype photograph reinforces the popular conception of better-off Irish immigrants before the 1840s Famine. (Courtesy Jeanne Baldwin and Vince Arnold, Museum of Newport Irish History)

these rough workers, hardening, even further, local prejudice and attitudes. As early as 1826, for example, a Newport newspaper reported a Sunday riot by several dozen intoxicated Irish laborers who worked on the fort. Other incidents in Rhode Island described drunken assaults, thefts, and even a bank robbery by "Irishmen." Occasionally the altercations pitted Irish immigrants from different counties against one another although one account blamed the arrogance of the more accepted Irish Protestants from Ulster for some of the troubles. (Many extant gravestones from this generation inscribe the provincial county of birth but not Ireland itself!) In another instance in Pawtucket in 1837 an inebriated Irish immigrant swam across the Blackstone River to take up the challenge of a fight with one of his countrymen who provoked him from the other side. He drowned before he could reach the other bank, prompting a local newspaper to comment wryly that no one anticipated "COLD WATER would be his ruin." Other Irish suffered frequent injuries, some fatal, from toiling at sometimes dangerous worksites. Several died in cave-ins, explosions, and rock falls. Infrequent descriptions of Irish gatherings might, back-handedly, remark on the unusual absence of intemperance or violence like the first local observance of St. Patrick's Day in 1839 at a banquet in Providence. Although some of the uncomplimentary incidents might be explained or analyzed away by the hardscrabble and alien life of these newcomers, the overall nature and frequency of the problems stick in historical accounts and contemporary mindsets. At least one honest analysis about the problems of work and the role of the Irish appears in a private letter in 1850: "two of my men have mutinied & gone off drunk. Colored brethren—but we have eight Irishmen & five natives left. How could the work of the country be done but for the Emerald Islanders."[15]

On the other hand, some Irish became victims of violence at the hands of the native-born. In August 1839, a brawl between two Yankee shipping masters on Providence's waterfront threatened to get out of hand. An Irishman named John Harman helped separate the two combatants. One of the fighters, Solomon Hicks, an influential Yankee, tracked down Harman an hour later and beat him to death for interfering in the quarrel. He allegedly bragged that "I have whipped that Irishman, and can sleep like a horse to-night."[16] Despite overwhelming evidence of premeditated murder, the Yankee jury, with encouragement from the state's Attorney General, convicted Hicks of manslaughter with a light sentence of only several months in jail. An anonymous letter writer lamented the leniency in an unusually sympathetic plea for reverse justice that still included the usual stereotype: "Now, Mr. Editor, I have yet to learn that, a hapless Irishman, whether firm or infirm;—whether drunk or sober;—whether peaceable or quarrelsome; whether surrounded by friends who are respectable or by friends who are not, is less entitled to the protection of the law, than the most elevated citizen in the land."[17]

Many individuals and a few groups of Irish managed to escape the victim/instigator category altogether. Michael Reddy, one of the earliest pre-Famine immigrants, arrived in Boston from County Carlow in October 1823. He spent three weeks there working as a laborer. Reddy quickly disliked city life with its culture of drinking and carousing. "He had left his home and friends to make his fortune not to squander it in riotous living," according to a local historian who interviewed him in 1874. Reddy walked all the way to Woonsocket and then on to Providence. The following spring, 1825, he met Edward Carrington along the Pawtucket Turnpike, "a meeting of the capitalist and the laborer." Carrington hired Reddy to work at his store in Providence but soon transferred him to a construction workforce on the Blackstone Canal, which eventually stretched from Providence to Worcester, Massachusetts. The artificial waterway paralleled and intersected with the Blackstone River, requiring the building of locks in some areas but very little work in other places. By the fall of 1826 Reddy had shoveled his way with other laborers to Woonsocket where he changed jobs. He stayed in town for the rest of his life except for a trip back to Ireland to find a bride who gave birth to seven children here. (Other Irish canal diggers went on to Worcester and formed that community's first Irish colony.) Ironically Reddy lived on a piece of property in Woonsocket that would eventually contain part of the stone foundation of Banigan's rubber company.[18]

Furthermore, Reddy invited a pioneer Catholic priest Rev. Robert D. Woodley to this area to hold mass for ten countrymen in 1828. They raised fifty dollars in expense money. Reddy also helped arrange the legendary visit of Father Theobald Mathew, the famous Irish temperance leader whose appearance in 1849 prompted local mill owners to suspend work so that operatives could attend the presentation. The religious teetotaler, in examining the books of Quaker relief funding for the Famine, announced that the Woonsocket Irish, proportionately, donated more than any other place in the United States. Local Catholics still celebrated the priest's visit as late as its fiftieth anniversary in 1899. Reddy served as an officer in the Father Mathew Total Abstinence Society, which dotted parishes throughout the British Isles and the United States in the second half of the nineteenth century. He also assisted with fundraising for the area's pioneering Catholic Church. When he died, the local newspaper described the crowd at the funeral: "a great number, too, of the American neighbors of the deceased were in attendance, as mourners for the loss of a good citizen, who, never forgetting the country of his birth, or the religion of his fathers, was ever true to the country of his adoption."[19]

There was another surprise in the uneven experiences and characteristics of the pre-Famine generation in Rhode Island. By the time the Banigan family arrived in Rhode Island in 1847, forerunners from County Monaghan had preceded them by thirteen years in a very different set of circumstances. Local manufacturers here recruited a group of skilled Irish bleachers from Ballybay, County Monaghan, for steady jobs in the state's cotton industry just as flax and linen employment flagged in Ulster and became industrialized outside of County Monaghan. Ballybay, less than ten miles away from the Banigan potato patch in Lisirrill on the Shirley estate, expatriated more than one hundred sons and daughters to work in the burgeoning mills of a textile community in Riverpoint, Warwick [currently West Warwick], Rhode Island. Here, they flourished with transferable textile skills and enjoyed longevity thousands of miles from any Famine affliction. They founded the oldest extant Catholic church building in the state, St. Mary's, after spending innumerable Sundays walking miles to the closest available services in Providence. A Catholic priest, the Reverend James Fitton of Boston, celebrated the first mass there in 1838 in temporary quarters. In 1844 the Irish immigrants built St. Mary's as a beacon to faithful, rural Catholics. A retrospective article in 1911 noted that "nearly all the men and women who left the flax fields in County Monaghan so long ago have found their resting place."[20]

Despite having been recruited for specific jobs and leaving Ireland voluntarily, the small community still suffered that almost genetic Hibernian sense of loss and exile. Anna Gibson, from County Monaghan, purchased a one-acre plot of land for St. Mary's church and cemetery. She wrote in 1859 that:

Oh! precious thoughts of home will start
At thy loved sound, sweet bell,
And many an exiled head bowed low
Beneath its weight of care
Will rise, forgetful of its woe,
And breathe the hopeful prayer.
Thou will recall the path he trod
On 'Erin's Emerald Isle,'
The glittering streams, the daisied spot,
The young heart free from guile.[21]

These rural Irish arrived in small groups, built tidy, bucolic homes, financed by steady, respected employment. The little outpost in River-point and several other suburban colonies initially posed no threat to the thin seam of Yankee employees there. On the other hand, a daughter of the original Irish settlers remembered hostility, her uncle being stoned by Yankee children in the village and derisively called "Paddy." The family lived on the appropriately named Ballybay Hill. The Banigans may have had some previous knowledge of Rhode Island through these immigrants from the same Irish county.[22]

Despite the uneven reception of the pre-Famine Irish by their Rhode Island hosts, many local authorities soon made scapegoats of these disparate newcomers in order to unify the citizenry against Irish-Catholic immigrants who might become naturalized and obtain the right to vote. Joseph Banigan's subsequent achievements cannot be appreciated without understanding the institutional and cultural impediments faced by Irish Catholics in Rhode Island—based on ancient religious prejudices and ethnic fears—which surpassed discrimination faced elsewhere. Yankee Rhode Island had imbibed most of the historical prejudices against Irish Catholics. Discrimination was directed at both sides of the Irish-Catholic hyphen. The attempt to disentangle religion from ethnicity has encouraged some academic studies, but to separate one from the other is like trying to separate Siamese Twins. Most local nativists twisted Irish and Catholic into a vituperative whole. The infallible Pope, conniving priests, and the Wild Irish already had a reputation long before Irish Catholics arrived in large numbers to feed a newly energized anxiety. The early arrivals seemed less put upon only because the subsequent Famine generation felt the full fury of reinforced discrimination.[23]

In 1830 the Reverend Francis Wayland, president of Brown University (more of a religious institution at the time than the secular giant of today) called the Catholic Church the "Scarlet woman of the Apocalypse" in a lecture at the school in Providence, although he employed an Irish coachman.[24] William Douglas, an evangelical Baptist minister and urban missionary to the poor, authored the annual reports of the Female Domestic Missionary Society in Rhode Island for decades. Beginning in 1833, he attacked Irish immigrants. He also cited several hundred local residents who practiced no religion at all and concluded that "these added to the Catholics . . . will after all swell the number of ignorant and vicious foreigners in this city to at least fifteen hundred. These persons usually form the nucleus of whatever is vile in association, and the leaders in whatever is abandoned, or daring in vice and outrage."[25] Thomas Man, an influential local writer who intelligently excoriated the faults and excesses of the factory system in one publication, spewed greater venom at black slaves and immigrants in another pamphlet in 1835. He wrote, "let England still continue to keep open the Irish flood-gate of moral pollution; (look at New York, during her elections, where Free Suffrage exists) inundating the Country with her vagrant and corrupt Catholic Priests—and their base and ignorant Convicts of the Roman Catholic Religion."[26]

The editor of the Providence City Directory in 1838 complained that "another difficulty in obtaining the names correctly arises from the ignorance of some of whom information is sought, they being totally unable to pronounce intelligently or spell correctly, their own name." He concluded by identifying the culprits: "This difficulty occurs mostly among the Irish part of our population."[27] Some other local Protestant missionaries, beside the usual lamentations against Irish Catholics, complained that illiteracy made the distribution of religious tracts among them a waste of time for purposes of conversion.[28] A teacher in the mill village of Natick in Warwick wrote privately in a letter in 1846 that "as my scholars were small I did not make a very powerful speech on the first day. Some of the little urchins . . . are as dirty as any hog—real Irish!"[29]

"Democracy in Decline"

Industrial unrest in the state preceded any social upheaval by the Irish. The employment of significant numbers of young children in the mills engendered some controversy, but the payment of hard cash during a period of primitive bartering helped assuage any parental concern by farmers. The strike by teenage girls at Slater's Mill in 1824 became a dress rehearsal for later labor actions that would pit rank and file Irish Americans even against Joseph Banigan after he became an industrial magnate in the postbellum era. This first "turnout," one of the initial factory protests by females in the new nation, seldom appeared on the state's list of firsts. Soon the arrival of newcomers from Ireland, and then French Canada, and elsewhere provided a steady and desperate group of "hands" more willing to tolerate low wages and long hours than native-born employees, especially as textile technology simplified the work and created even more unskilled jobs.[30]

Artisans in the state had already organized themselves in 1789 into fairly sophisticated unions even before the establishment of Slater's Mill. The Association of Mechanics and Manufacturers formed two branches in Newport and Providence to promote general interests, shorter workdays, and a welfare fund for members. The group embraced public education and temperance, and lobbied for an end to voting restrictions, the latter stand placing these native-born tradesmen on the cutting edge of the forthcoming Dorr War in 1842. Carpenters, mechanics, printers, and other skilled workers found it difficult to proselytize the younger and less skilled operatives in the state's many textile mills. The appearance of similar primitive labor unions throughout the East in the 1830s, during the era of the workingmen's political parties, failed to make any serious inroads among unskilled, recent Irish arrivals. The Panic of 1837 led to the financial crash of many factories and unglued the hard work of these early trade unionists who could not cope with the economic depression any more than the owners. However, the idea of worker groups stayed in the minds of many laboring people until a more propitious time. In Rhode Island much of the energy went into agitation around the ideals of the approaching Dorr War.[31]

During the formative years of the factory system in Rhode Island the early wave of Irish immigrants failed to gain widespread entrance to the world of textile production. In 1828, when the Yankee Wilkinsons announced the gift of land in Pawtucket for St. Mary's Catholic Church, several newspapers applauded the open-mindedness and generosity of the family as well as the beneficial and moral impact that a regular place of worship would have on the many Irish newcomers toiling in the mills. Several letters to the editor challenged the assumption that the Irish worked in the mills. The editors of one publication soon apologized.

> We were then under the impression that a "considerable portion" of the persons employed in the various establishments at Pawtucket were foreigners, and probably Catholics. . . . We . . . learn that this is a mistake, and that of the large number of persons engaged in Manufacturing establishments, but very few, if any are foreigners. There are some Irish, employed in hand weaving in the village, and others who work about the factories and elsewhere. . . . The fact that so few foreigners are employed in the Factories in Pawtucket, is highly creditable to native skill and industry.

Only a generation later local historian Charles Carroll stated positively that the Irish "were strongly entrenched in factories."[32]

Although immigrant labor eventually replaced the bottom-rung workers in the state's ballooning textile empire, the changeover was gradual and many native-born employees climbed the ladder of success in cloth production by achieving skilled and managerial positions. During those first few generations of factory expansion and the concomitant growth of adult employment in related industries like machine and tool production, Rhode Island became a beehive of industrial activity. As the workforce in this arena grew, the voting rolls diminished as most toilers left the family farm or arrived from foreign shores with more things in their memories than in their luggage. By 1840, the land of Roger Williams had degenerated into a "democracy in decline." Almost 60 percent of the adult, native-born, white male population could not cast a ballot because of a lack of landed property and sufficient tax payments. The primitive apportionment system of two representatives for each new town paralleled the drop-off in political eligibility. Shrinking rural populations with a handful of voters (a few not much larger than Banigan's townland of Lisirrill), almost like English rotten boroughs, enjoyed as much representation in the General Assembly as emerging industrial cities with burgeoning populations. Factory-laden Smithfield, for example, harbored more than 9,000 residents in 1840; Barrington hosted fewer than 600. Both sent two delegates to the General Assembly. The King Charles Charter of 1663 had not anticipated the machine in the garden or its consequences.[33]

Another egregious shortcoming of the old charter was the lack of a specific mechanism to amend outdated clauses or add required provisions. Only the General Assembly could initiate constitutional reform through a constitutional convention and, as the urban, industrial forces expanded, the old guard agricultural areas impeded any changes that would surely diminish their land-based and now undemocratic power. Although the clamor for these alterations increased in volume, the legislature simply refused to consider a such an assemblage. By 1824, the year of the strike at Slater's Mill, the hue and cry for modification spread among a working class hungering for the vote. Similarly, their employers sought greater political leverage to match their tax contributions in the underrepresented cities. The voice of the state's past, steeped in bucolic village town meetings, spoke out against any liberalization that would share power with the intemperate, modernizing forces of manufacturing and Mammon. They feared, not unrealistically, that the wealthy mill owners might dictate to their employees to vote in a bloc without the protection of a secret ballot. The ever-shrinking pool of freemen that still possessed enough property in the form of real estate to reach the threshold for voting eligibility, simply chose to maintain their influence and the ways of the past. Those eligible to cast a ballot refused to voluntarily share the franchise with other, poorer citizens. Two constitutional conventions, in 1824 and 1834, purposefully accomplished nothing, more stillborn than alive. Nevertheless, the rhetoric around the question increased.[34]

The tension simmered for almost a generation. One articulate opponent of change threw bitter accusations at the reformers: "the right of suffrage . . . is put into the hands of that class of people . . . who have little or nothing at stake themselves; care little or nothing for the rights of others—people who in voting exercise no judgment of their own nor have any wish to form any; taking their impulses and directions from their leaders and ready to fight their battles, not merely by voting, but by brawling and violence."[35] More pointedly, a broadside signed "Roger Williams," aimed at "American Citizens," warned of the baneful immigrant influence infecting Rhode Island: "Bands of filthy wretches, whose every touch was offensive to a decent man, drunken loafers, scoundrels who the police and criminal courts would be ashamed to receive in their walls. . . . We saw Irish priests there—sly, false, deceitful villains looking on and evidently encouraging the gang who started the tumult."[36]

These sentiments targeted not only Irish-Catholic immigrants but native-born artisans as well. The restive and active members of several chapters of organized, skilled workers fought back against the powers-that-be, agitating for full political rights through speeches, petitions, and pamphlets. The conservative forces, the influential defenders of the past, encountered some difficulty in making their unwarranted attacks against the employed, semiskilled, and skilled workers of Yankee background. These artisans paid taxes, though not always enough to reach the voting threshold, and fulfilled a variety of civil service requirements in the militia, fire brigades, and the upkeep of public property. Tradesmen, who in many ways were the backbone of society at the time, made poor bogeymen for those seeking to maintain the status quo. They faced a difficult challenge: dueling the agricultural powerbrokers for the right to vote while, at the same time, engaging employers who controlled wages and working conditions but who remained potential allies over the other constitutional issue of representation in the General Assembly. The defenders of the status quo feared the growing numbers of urban workers, skilled and unskilled, and the message of equal justice that also resonated among other citizens not part of the "dirty apron" crowd. By 1840 approximately 17,000 toiled in the state's couple of hundred factories. The new working class lived predominantly in the five largest municipalities, comprising almost half of Rhode Island's population but represented by only a handful of political delegates compared to smaller rural areas. On paper, at least, native-born workers did not appear to be opposed to equal rights for the Irish in a battle similar to their own struggle for most of the same things. About the same time approximately 1,700 Irish lived among the 25,000 inhabitants of Providence while the state harbored 5,180 Hibernians (more than lived in neighboring Connecticut) in a population over 110,000, according to diocesan surveys.[37]

The Dorr War

The ferment that began in the 1820s against Rhode Island's restrictive voting and uneven apportionment continued to brew. Thomas Wilson Dorr, a blueblood aristocrat, emerged to bring the agitation to a boil. The scion of old Yankee families, legally trained and politically active, Dorr possessed a burning desire to restart the state's lost democratic fervor by embracing universal manhood suffrage. Dorr and his followers, a motley crew of fellow patricians, artisans, and immigrants, boldly bypassed the King Charles Charter of 1663 and the General Assembly. Dorr authored a "People's Constitution" endorsed by an extralegal People's Convention in the fall of 1841. The progressive document included clauses liberalizing voting rights and representation in the General Assembly, as well as provisions for a secret ballot, public education, a state bill of rights, and a streamlined amendment procedure to provide future redress. The document retained some voting restrictions for financial referenda and failed to enfranchise people of color, despite Dorr's opinion to the contrary—perhaps from a fear that the scale of reform would seem too Jacobin. Ironically some Dorrites descended to racist ridicule when the black community responded to a conservative initiative and later served in the state militia, thus securing their own right to vote. Most decisively, Dorr's blockbuster document did provide the ballot to poor, native-born, and naturalized immigrant males irrespective of affluence in most political contests.[38]

The reformists held a plebiscite, bordering on insurrection, in December 1841. It was open to current freeholders and those eligible to vote under the provisions of the new Dorrite constitution but not according to the old rules. In an election scrupulously administered, the results of which were later printed in a report of a United States Congressional inquiry, voters overwhelmingly passed the progressive measures 13,944 to 52, a majority of both sets of the electorate. In Providence, for example, 1,060 qualified freeholders cast ballots along with 2,496 formerly unqualified voters, expanding the electorate tremendously. Obviously the small handful of negative votes suggests a boycott by some freeholders, but a majority of eligible as well as ineligible voters still endorsed the monumental changes, at least in the opening chapter of this event.[39]

A cursory examination of the voters included in the 148-page list in the Congressional report yields only a smattering of Irish names, though the forces of Law and Order had mobilized around the myth that the results of the election would surrender the state to Hibernians. They could of course have had no knowledge of the impending potato blight and Famine exodus. Conservative Democrat Elizha R. Potter Jr., representing an agricultural area of the state, remarked at the time that his neighbors "would rather have the Negroes vote than the d[amn]d Irish."[40] The *Providence Journal,* in a fusillade of editorials by Henry Bowen Anthony before and after the 1842 Dorr War, employed verbal grapeshot to shred the possibility of civil and political rights for Irish Catholics. One representative opinion piece claimed "Rhode Island will no longer be Rhode Island when that is done. It will become a province of Ireland; St. Patrick will take the place of Roger Williams and the shamrock will supercede [*sic*] the anchor and Hope [the state's motto]."[41] One reformer lamented that "Men were called upon not to vote for a constitution but to vote against Irishmen."[42] The Law and Order Party used the Irish-Catholic "question" to obfuscate the legitimate need for change. Understandably the Rhode Island Irish chose a low profile in the ensuing "war."[43] A former governor of the state wrote in a private letter that said "my Irishman Patrick who was boiling over with fight came home from Chapel . . . and said that the Irish were to take no part in the quarrel, Father [John Corry] having interdicted them. This story has been confirmed from many quarters. The Bishop put his injunction upon them."[44] Later the conservative government instituted a dragnet to arrest several hundred Dorr supporters. In newspaper compilations of the arrested, only a couple of Irishmen appear. At least some Irish parishioners resented the pacifist yet pragmatic stand taken by the Catholic Church, but few would actually join the upcoming fray. However, many still had one eye cocked on political and nationalist developments in the homeland.[45]

The verbal jousting, before and after the Dorr War, provided more fireworks than any military action. After the victorious referendum in December 1841 to change the constitution, the ruling Whig Party transformed itself into the Law and Order Party by preparing for what seemed like an inevitable clash in many arenas. They mustered the judicial system, the political process, and the military to clamp down on the protestors, using a draconian edict called the Algerine Law by progressives. They also started an initiative to undercut the reformers by offering their own constitutional changes, changes they had resisted for a quarter of a century until the opposition took action. The Law and Order advocates proffered the right to vote to white, male, native-born citizens without a property qualification, but they cagily kept the tax requirement for citizens of foreign birth. The Dorrites, perhaps lulled into a false complacency by the seeming popularity of their cause, fiddled into 1842 before attempting to consolidate their stunning triumph. By April, the state had two separate governments and governors. Push had come to shove.[46]

A botched nighttime raid on the state arsenal by Dorr's amateur forces ignited the insurrection in May 1842. The Law and Order governor unsuccessfully sought federal troops from President John Tyler. Meanwhile Dorr visited Tammany Hall in New York City and received a pledge that an army of Irish Americans and other Democrats would invade Rhode Island, one of the few times any evidence surfaced of such outside activity. Only a handful of residents of the notorious Five Points section of New York ever appeared, but even the rumor was enough to convince many citizens that the threat was real. A Law and Order poster ridiculed the invasion with simian caricatures of Irish volunteers. The conservative government mobilized the state militia and, in one of the most Faustian deals imaginable, promised to enfranchise African Americans in the state who they had disenfranchised in 1822 in exchange for their physical support against Dorr. Still stung by the reformers' abandonment of the black community, people of color seized the opportunity to join the Law and Order troops and secure the ballot through the bullet.[47]

A subsequent "battle," pathetic in nature, led to the arrest of several hundred Dorrites, with two accidental deaths, and a short reign of terror in the state by champions of Law and Order. The conservative conquerors followed through on the promise to open the ballot to native-born white and black males but restricted the Irish and later immigrants to second-class status by maintaining a landowning requirement for foreign-born voters. Their constitution passed by as significant a margin as the People's charter just a year earlier, because it was boycotted by the reformers. The revised charter failed to provide secret ballots or a manageable amendment system. The conservatives did institute a partial reapportionment plan, providing more representatives in the

General Assembly to urban, industrial areas. The legislature's real power remained in the state senate controlled by rural interests. That authority lasted, almost immutably, until the "Bloodless Revolution" in 1935 when a New Deal coalition of ethnic Democrats finally toppled the aged Republican political machine. Whatever the shortcomings of the new statutes, the reformers really deserve the credit for the changes that conservatives had resisted but finally made when pushed into action. Dorr, although less famous, is now perceived as a guardian of the legacy of the colony's founder, Roger Williams, who was cited in the reformers' constitution.[48]

Dorr, who sought refuge in several nearby states sympathetic to his cause after the upheaval, decided to surrender to Rhode Island authorities on 31 October 1843 and face a trial for treason. That same year on the Shirley Estate in County Monaghan, Ireland, the struggling peasants and farmers had tumultuously demonstrated against land rents in March and physically challenged the authorities, who incited opposition when they posted legal debt notices on the doors of Catholic churches in June. Joseph Banigan, a toddler at the time, probably learned of these struggles from his family and compatriots after his arrival in Rhode Island several years later. Ironically, the land and voting issues, although somewhat different in nature, must have struck Irish-Catholic refugees from the Potato Famine who came to Rhode Island in the late 1840s.

The Sprague Murder and the Gordon Brothers' Trial

Although the direct and indirect ramifications of the Dorr War held sway in Rhode Island well into the twentieth century, another event on the last day of 1843 created an even more hostile atmosphere for Irish-Catholic immigrants. On 31 December 1843, New Year's Eve, someone bludgeoned to death one of Rhode Island's most prominent industrialists, Amasa Sprague. Local authorities charged three Irish-Catholic brothers named Gordon as the culprits. A set of trials in 1844 and 1845, about the same time the courts sentenced Dorr, constituted the most spectacular legal proceedings in Rhode Island history and provided new fodder for discrimination.

The Sprague family owned the A&W Sprague Printing Company in Cranston, next door to Providence. They manufactured calico cloth by impressing designs onto cotton textiles. William Sprague, the older brother, served as Rhode Island's United States Senator in Washington while Amasa ran the plant on a daily basis. A number of Irish immigrants labored at the facility among a workforce of several hundred. In 1836 an Irish-Catholic newcomer, Nicholas Gordon, probably from County Tipperary, opened a small store near the mill and slowly expanded it into a tavern with a local liquor license, a common enough evolution into a community center not unlike bars in Ireland. Over time, many textile employees walked the short distance to have lunch and an alcoholic drink before returning to work. Amasa Sprague, concerned about the condition of his employees and probably fearing potential accidents or incidents, successfully petitioned the Cranston Town Council in person to encourage the cancellation of Gordon's liquor permit in July 1843. Gordon, who made a good living from the bar, seemingly had qualified to vote because he owned enough property to meet the franchise threshold, an unusual circumstance for an Irish-Catholic immigrant in the era of the Dorr War. He had attended the inaugural Saint Patrick's Day celebration at a banquet in Providence in 1839, making a public toast to a New York newspaper editor as being "always vigilant in defending Irishmen against the attacks of the zealot and bigot." The words would have a mortal, personal application in several years. Ironically, Amasa Sprague and other significant employers of Irish immigrants were invited to the same event. Sprague did not attend, but the family initially favored Dorr's reforms until embracing the Law and Order side, probably because of the violent turn of events. The Sprague Mansion museum contains a silver set engraved to the editor of the *Providence Journal*, Henry Bowen Anthony, for his support of the status quo.[49]

Reflecting his relative affluence, the politically active Nicholas Gordon sent for his extended family (three brothers, a sister, niece, and his mother) as so many immigrants did before and after him. They came in the summer of 1843, during the fallout from the Dorr War and shortly after Amasa Sprague had the liquor license canceled. Although the journey must have taken months to arrange properly, the arrival so soon after the withdrawal of the permit seemed more than coincidental in the popular mind, still inflamed by the recent constitutional troubles.[50]

When Amasa Sprague's disfigured body was discovered with his personal valuables untouched, the crime seemed to reek of hatred rather than robbery. The local posse headed directly to the nearby Gordon homestead following just one of many sets of prints in the snow. The authorities arrested the four brothers, the mother, and the family dog. The same state attorneys who prosecuted Dorr would also charge the Gordons with conspiracy to murder Sprague in retaliation over the alcohol license issue. The interrogators tried William and John Gordon first, in April 1844. Nicholas, the older brother who brought them to Rhode Island, would have a separate trial later as the alleged ringleader. The fourth brother was not charged. The entire case rested on circumstantial evidence as no one actually witnessed the crime. Job Durfee, the chief justice of the Rhode Island Supreme Court, in his instructions to the jury about the murder, declared that "it has no parallel in the annals of the State, nor one which can exceed it in the annals of any one of the United States." Irish-Protestant former Governor John Brown Francis, who attended Sprague's funeral and urged Senator William Sprague to take control of the prosecution, would write that "from what I hear of the murder case there is very little prospect of a conviction. . . . I doubt whether you can get a Dorrite in the county of Providence to consent to hang a man for doing what they will consider rather a venal offence."[51]

The trial revolved around physical evidence from the crime scene, and indeed some of it seemed to implicate the Gordons, especially a gun used in the murder, which apparently belonged to Nicholas. However, other aspects of the case raised serious questions about the impartiality of the investigation and the proceedings. Like all such trials, this one was full of loose ends. But the fiber of anti-Irish-Catholic feeling, so powerfully exploited during the Dorr War and further nurtured during Dorr's almost concurrent hearing, ran like an invisible thread through both hearings. The local Irish population chipped in to finance the defense effort. One of the state's lawyers linked the revenge conspiracy to family solidarity. "The tie of kindred is to an Irishman," he told the jury, "almost an indissoluble bond." Nicholas Gordon, according to the prosecutor, bragged about Ireland and the Irish. His brothers, so respectful of him, found "a pleasure in fulfilling his wishes and advancing his plans."[52] Even the defense team led by Thomas Carpenter employed ethnicity in one of the subsequent trials: "There may be those among you who think there is little difference between taking the life of a dog and that of an Irishman."[53] In the summation, the main defense attorney pleaded: "Do you suppose the prisoners are differently constituted from other men—that because they happen to be Irishmen, they are not made in God's image, and endued with human natures?"[54] In his instructions to the jury, the chief justice told the talismen to give greater credibility to the testimony of Yankee witnesses than that of Irish-Catholics.[55]

After a nine-day inquiry the jury returned the verdict in seventy-five minutes: John Gordon was guilty but his brother William, who had a decent alibi, was acquitted. John was sentenced to die the following February 1845. In the meantime, Nicholas Gordon, the oldest brother and proprietor of the ill-fated tavern, faced the same murder conspiracy charge in October 1844 in the same prejudicial atmosphere. Prosecutors portrayed him as the mastermind of the vengeful murder of Amasa Sprague in a replay of the first tribunal. After sixteen hours of sifting the testimony of one hundred witnesses, the jury returned a shocking split decision: eight for conviction, four against. The indecision led to another trial in April 1845, two months after William's brother John was to hang as a conspirator in the slaying. With only one guilty verdict, a conspiracy did not exist legally. The defense team called for a postponement of John Gordon's scheduled execution in February 1845 until after the second trial of Nicholas two months later in April. The governor, arch-conservative Democrat James Fenner, begged out of making a decision, citing a technicality, and threw the judgment to the General Assembly, which voted thirty-six to twenty-seven in a jolting decision to proceed with the hanging on time.[56]

John Gordon, the twenty-nine-year-old Irish immigrant, walked to the gallows in the state prison yard on Valentine's Day, 1845. Thomas Wilson Dorr, recently sentenced to life imprisonment for treason, probably watched from his tiny cell. The authorities invited sixty Rhode Island notables to witness the execution from within the facility. Former Governor John Brown Francis suggested repealing an 1832 law requiring private executions because "a good hanging in the presence of a mighty assemblage might have a salutary effect." One thousand observers, presumably many Irish Catholics, stood outside the perimeter walls. Father John Brady, a Catholic priest who traveled the New England circuit, heard Gordon's confession and accompanied him to the scaffold. As the hangman slipped the noose over the head of John Gordon, Father Brady made a loud pronouncement that shocked the elite onlookers. "Have courage, John." he bellowed. "You are going to appear before a just and merciful judge. You are going to join myriads of your countrymen, who, like you, were sacrificed at the shrine of bigotry and prejudice. Forgive your enemies."[57]

Would the priest have made such an inflammatory statement after hearing John Gordon's penultimate confession? Or did the young Irish immigrant, in a strange land amidst a bewildering controversy, lie to Father Brady? It is hard to imagine that John Gordon, at the threshold of death, abandoned his only hope of salvation through divine forgiveness by deceiving the priest. Would Father Brady have uttered his contentious statement if Gordon had actually admitted his guilt? Twenty minutes later all the questions became academic. John Gordon was dead and, unbeknownst to anyone at the time, he would be the last person executed by the state of Rhode Island. The massive funeral cortege marched boldly by the Sprague Mansion and detoured in Providence to pass by the scene of the trial in one last collective mocking of the justice system by the local Irish.[58]

Two months later, in April 1845, prosecutors tried Nicholas Gordon a second time. The examination rehashed familiar arguments but in a different political atmosphere. The life sentence of Dorr also took a toll on the public conscience and a "Liberation Party" temporarily broke the political grip of the Law and Order party in the spring elections. Nicholas Gordon's second trial ended in another sensational mistrial: nine voted to acquit, only three to convict. The state lost the will for another go-around at about the time the potato blight made its first appearance in Ireland in the fall of 1845. John Gordon was hanged as a conspirator without any convicted accomplices. The psychological blight of the trials helped eventually to end the death penalty in Rhode Island, in 1852. Eighteen months after his last court appearance, Nicholas Gordon died of natural causes. Two academic writers concluded the only full-length study of the Gordon murder proceedings in 1993 by writing that, "from the perspective of nearly a century and a half, it is apparent that John Gordon was a victim of widespread prejudice against the Irish."[59] The authors went on—without even as much circumstantial evidence as the prosecution in the trials—to suggest that William Sprague, the U.S. Senator and

Amasa's older brother, had greater business motives to get his brother out of the way![60]

No Catholics served on the three Gordon juries; the judge upheld every challenge by the prosecution and denied all of those from the defense; the state Supreme Court acted as its own appellate body in the case; and the governor and the legislature, for all practical purposes, voted to hang John Gordon before Nicholas's second trial and the establishment of a conspiracy. Defendants in capital cases could not even testify at the time. The posse followed only one set of tracks near the murder scene and ignored dozens of others. The residence of the chief prosecution witness, a prostitute, was owned by the brother of one of the sitting judges on the Supreme Court, who had been sued for cheating the Flemings, a family of Irish immigrants, in arranging passage to Rhode Island. Furthermore, local newspapers lined up almost exclusively against the Gordons, often leaving out any favorable evidence or arguments in favor of the defendants. The press in other places, especially where the Irish enjoyed some political power, provided a different take on the event. The *Baltimore Republican,* for example, called the hanging "a judicial murder—alone worthy of the State in which it was consummated."[61]

Nevertheless, the circumstantial evidence—so easy to belittle—stained the defendants. A fowling piece used to bludgeon Amasa Sprague seemed to belong to Nicholas Gordon. And the Yankee juries, required to possess the franchise in order to sit in judgment, carefully—to their credit—convicted only one of the three brothers. But the truth will never be known, only the prejudice of the times. Gordon's mother lost the family home and business, heavily mortgaged from liens to raise money for the trials. She lacked $1.20 in taxes and the building sold for $4.75 in 1851. Rhode Island historian Joseph Sullivan has uncovered other possible defendants with as much if not more motive to kill Amasa Sprague as the Gordon family; but these potential suspects do not appear in the literature of the trials or newspaper accounts of the times. Henry Mann, author of a later history of the Providence Police Department, wrote that a model of the Gordon gallows and the remnants of the actual noose stood on display in the office of the state's attorney general. "The rope was rapidly vanishing," he noted, "many old Irish women begging pieces of it for charms."[62]

The Yankee Ascendancy

Between the execution of John Gordon and the second trial of Nicholas Gordon, Charles Jackson, a new governor of Irish Protestant background, freed Dorr on 27 June 1845. Almost a decade later the General Assembly annulled his conviction, just before his death in 1854 at the age of forty-nine. The Law and Order forces, temporarily out of power from a backlash against the haunting Gordon trials and the life sentence of Dorr, issued a broadside in 1846 claiming that Jackson also planned to make it easier for immigrants to vote. The flyer, entitled "Chas. Jackson and Irish Voters," claimed the liberal governor wanted to enfranchise Catholics in order to allegedly "OFFSET THE IRISHMEN AGAINST THE NIGGERS." The anonymous author stated that farmers would "become the slaves of Roman Catholic tyranny! The Pope of Rome has but to issue his mandate to the priests, and those priests will ever after control all your elections."[63]

The pre-Famine settlers in Rhode Island, still numbering only several thousand, congregated in Providence, especially in the Fifth Ward, the Fox Point section of the city, and around Philip Allen's mill in the North End. They displayed considerable resiliency in agitating for civil rights after the Dorr War and Gordon trials. In a petition to the state General Assembly in 1846, the same year as the incendiary handout and the Banigans' exploratory journey to Scotland, Irish Catholics led by Henry Duff complained of the inequity of being residents without the right to cast a ballot. They called it "a degrading brand placed upon all naturalized citizens, setting them apart as a peculiar, unworthy and disqualified class, to be shackled and held in suspicion, and continual objects of insult and derision."[64] They further remonstrated to Congress that they had been degraded below the status of former slaves. Animosity against Irish Catholics in the United States manifested a virulent strain in most areas of the country, but the Dorr War and the Sprague murder created a disease unmatched elsewhere that no amount of later second-guessing could cure.[65]

The political status quo boomeranged between Whigs and Democrats after the liberation of Dorr. In fact, Dorr's uncle, the liberal industrialist Philip Allen, became governor in 1851 for several one-year terms just as Irish Catholics created a fledgling parochial school system. Yet discrimination entered a more subtle and insidious phase.

The retention of the property requirement, more than the two-year residency clause in the conservative constitution, hampered the civil rights of immigrants. Famine refugees, who made up a majority of foreign newcomers to the state, suffered. They had faced two ascendancies in their lives: the Protestant one in Ireland and the Yankee one in Rhode Island. Still, the economic opportunities in the state trumped all the hardships and, if it proved to be a hardscrabble existence here, anything seemed better than the Famine.

During the first Gordon trial the chief defense lawyer Thomas Carpenter had explained the arrival of the brothers: "They came over from Ireland a few months since as hundreds of their countrymen annually do, to seek in our more favored land the employment which is denied them in their own."[66] The Banigan family, and thousands like them who settled in Rhode Island, must have felt very much at home in a kind of black comedy. They faced prejudice at the hands of the Yankees as thick as any Gaelic brogue in Ireland. Like the forty-shilling property requirement in Eire, the $134 threshold here severely limited political engagement, although the children of immigrants automatically became American citizens if born here. As in Ireland, there was no secret ballot until 1889, at a time when most eastern states had instituted that protective procedure. Without the right to vote, no Rhode Island citizen could serve on a jury, own property, or bring civil actions in a court. Among the pre-Famine generation in the state only a small minority ever qualified to purchase land directly, and the Catholic Church faced similar difficulties in buying real estate as well. In Rhode Island realty agents feared a local backlash against such deals, so Catholics, at times, relied on the kind disposition of a handful of sympathetic Protestants as straw buyers.[67] In Ireland, property restrictions were even more ironclad against Catholics. Rhode Island also retained laws for debt imprisonment until 1843 as well as requirements for militia and fire brigade service. The apportionment system, despite the tepid changes in the Law and Order constitution, insured a stranglehold of power for rural interests in the state senate not unlike rotten borough arrangements in the British Isles. All in all, the two trans-Atlantic ascendancies, both Protestant and Yankee, seriously curtailed the rights of Irish-Catholics.[68]

Although the immigrants faced a stacked deck in Rhode Island, a few sympathetic wild cards balanced the discriminatory trend. The irascible but honorable Roger Williams invited practitioners of the Catholic faith to the state during a colonial period of intense anti-Catholic hostility within the British Empire. Later, in the 1830s, the city fathers of Providence allowed them temporarily to worship in the Town House. Quakers around the state also aided Catholics in obtaining places to hold masses and later assisted in Famine relief efforts. The Episcopalian Wilkinson family in 1828 provided land for the earliest Catholic Church and cemetery in Pawtucket where John Gordon was probably buried after his execution in 1845, at a time when most Catholic individuals and organizations could not own land. A descendant of the Wilkinsons, as noted earlier, wrote a stinging missive against the Irish in 1901 without mentioning earlier toleration in his own family.[69]

Even a young, conservative Thomas Wilson Dorr had actually embraced discriminatory property qualifications against immigrants before rethinking his position and pursuing voting privileges for Irish-Catholic newcomers. He respected their religion and said so more than a decade before the "War" that bore his name. Echoing the toleration of Roger Williams, he declared; "If men cannot agree in religious opinions—and from the constitution of the human mind such an agreement can never exist—they certainly can agree to differ peaceably."[70] Thomas Carpenter, the Democratic candidate for Governor in 1842 and chief legal counsel for Dorr and the Gordons, later underwent a highly unusual, almost inflammatory conversion to Catholicism in Yankee Rhode Island. Even the maligned juries in the Gordon trial convicted only one brother in a prejudiced, volatile atmosphere. Protestant Governor Philip Allen (1851–1853), Dorr's uncle, aided the construction of the hallmark church (later the cathedral) of Sts. Peter and Paul in Providence. He also endorsed voting rights for naturalized citizens, several hundred of whom worked at his local print works, an unusual bromide to the usual No Irish Need Apply dictum. And, just as some ordinary citizens held racist views toward the Irish, there were obviously some who harbored a more tolerant view. Although these noteworthy groups and individuals could not stem the larger discriminatory tide, their sensitivity and assistance courageously countered the temper of the times.[71]

THE RHODE ISLAND IRISH

"They are not forgotten in their affliction."
—Woonsocket Irish [Famine] Relief Committee
upon collecting almost $3,500 in 1847

"Yesterday afternoon, Patrick O'Some-thing-or-other, we don't remember what, and it's certainly of no consequence, called at the Police office to make complaint against another Patrick whose name was Michael, saying that the last mentioned individual had knocked him down."
—Providence Post, *21 May 1859*

During the mass arrival of Famine refugees, the Irish population in Rhode Island swelled to almost 16,000 in 1855, constituting about one-tenth of the state's population of 150,000, and more than a quarter of the residents of Providence. Despite the reinforcements, the sense of banishment not only failed to dissipate, but the immigration experience took on a life and legend of its own as newcomers retold and enriched the reasons for their exile. As the situation deteriorated in Ireland, even organizations unsympathetic to Irish Catholics showed compassion, and the inclusion of Scotland in the relief efforts may have liberalized local Protestant aid and mindsets. The Quakers assisted, as they often did during humanitarian crises, regardless of religious denomination. As the Famine stalked their former homeland, earlier Irish arrivals donated relief funds and goods while honing their organizational skills. In February 1847, not many months before the Banigan family embarked to these shores, residents of Woonsocket collected over $3,000 for relief; $600 came from Irish workers in that city's many factories who wanted their brethren to know "they are not forgotten in their affliction."[1] Woonsocket, according to one historian, displayed the most ecumenism of any town in Rhode Island. Others eschewed organizational donations and sent small sums directly to affected relatives and friends. Father Mathew would compliment the generosity of the Irish in the Blackstone Valley during his trip in 1849. A generation later Joseph Banigan would employ hundreds of his countrymen there in the most dynamic rubber footwear company in the nation. He continued to amplify a practice of public and parochial philanthropy unprecedented in the state's history, building on an unusual charitable legacy of the Famine generation that still marks Irish generosity.[2]

Providence, harboring the lion's share of Irish Catholics in the state because of employment opportunities in the area, also practiced a charity and concern not publicly evident during the Dorr-Gordon trials. Even some antagonists of Irish Catholicism joined in the effort, citing Christian principles. With a smidgen of opportunism and religious guilt, some Protestants hoped that their benevolent efforts would provide the wherewithal to keep the Irish at home in Eire. Still their numbers continued to grow here, but the concentration offered little succor to the immigrants. As Rhode Island historian Thomas Bicknell wrote: "Their poverty, behaviour, speech and religion marked them as strangers."[3]

Banigan at Work

Banigan attended public elementary school for a year before going to work full time, an honored and necessary tradition for generations of immigrants who viewed the family as an economic unit. The nine-year-old Banigan entered the workforce in 1848 as part of a Rhode Island system that hired children in large numbers, harking back to the earliest days of Slater's Mill. He spent several years at the New England Screw Company, which merged with a similar enterprise to form the American Screw Company. The new entity, capitalized at a record $1,000,000, came to be one of the state's "five industrial wonders of the world." The enlarged enterprise employed 600 "hands" at its inception in 1850, half of them women.[4]

Seth Hewitt, an early labor organizer for the Fall River, Massachusetts, Mechanics Association, made a stop in nearby Providence in

1844. He wrote in his diary: "I am told on good authority, that one of the Screw Factories here, employs about forty children between the ages of six and ten through the long hours of the day, giving them no time to live! Here they are compelled to grow up dwarfs in body and dwarfs in mind. Is not this slavery?" He further wrote that

> methinks this screw factory is rightly named, for the very life and energy as those designed for living beings, are really SCREWED out of them. These children are mostly Irish, and I am told, the fact of their oppression having come to the ear of the Catholic priest, he advised the parents to take their children away as soon as they could possibly find other and more comfortable situations for them. That priest has some soul! which certainly cannot be said of all.[5]

Although industrialized children had an opportunity to find employment elsewhere in the period between the economic downturns of 1837 and 1857, the chance of discovering some sort of alternative, wholesome form of occupation was minimal. Exploitive owners might have mitigated conditions temporarily or simply replaced any immigrant children bold enough to leave. Certainly, the arrival of families like the Banigans provided an almost unending supply of desperate, destitute labor that lasted virtually uninterrupted through World War I. Banigan, even before his teenage years, witnessed and experienced the brutal industrial conditions that pockmarked the state. He must have developed a tremendous distaste for such circumstances that, combined with the family's travail on the Shirley estate in County Monaghan, inculcated an aversion to the exploitation of the young in corporate or agricultural settings. Seth Hewitt, the union agitator who toured Rhode Island in that era, gingerly crossed the line between labor activism and political enterprise even though he eschewed any electoral activity. A Rhode Islander told him that his labor movement would conjure up a "second edition of Dorrism" in the popular conscience despite its purely moral and economic objectives. Banigan would, upon his own industrial ascension, try to institute a Catholic morality and righteousness in labor-management relations, perhaps as a result of his childhood workdays. He hired very few children at his own footwear facilities. During his lifetime he bankrolled the construction of orphanages, working homes, and other facilities to assist juveniles. He donated at least $1,000 to the secular Woonsocket Day

Nursery. In his will he left $10,000 to the Rhode Island Society for the Prevention of Cruelty to Children and an equal amount to the Lying-in Hospital for childbirth.[6]

The youthful Banigan, after several years of toil at the screw and nail company, served a three-year apprenticeship in the jewelry business from about 1857, the year of a severe economic downturn, to 1860 when he turned twenty-one. The city hosted dozens of such specialty establishments, collectively employing over one thousand. Welcome Arnold Greene, a popular local historian, wrote about jewelry manufacturing in this period: "the right man in the right place could almost name his own wages."[7] Earlier most apprenticeships lasted seven years, but as time passed the industry turned to mechanization and unskilled labor, including the hiring of some Irish Famine refugees. Banigan might have worked at Claflin & Company, which specialized in the production of shell jewelry. He could have also toiled at the firm of Sackett and Davis owned by Dublin-born Thomas Davis, who represented Rhode Island in the U.S. Congress from 1853 to 1855. Davis had served an apprenticeship himself in the jewelry trade and, although an Irish Protestant, he spoke out against religious discrimination and might have embraced someone like the spunky Banigan. Historian Charles Carroll wrote that some Yankees "were more than merely generous in helping ambitious Irish boys and girls to progress." As job opportunities expanded in some industries Irish immigrants sought employment and elevated the population of Providence from forty thousand to fifty thousand between 1850 and 1860, ballooning the percentage of Irish parents and children to a third of the city's numbers.[8]

Banigan liked to "tinker," a word often associated in that era with Yankee inventors and artisans, not Irish-Catholic immigrants. In fact, an article about the Claflin enterprise mentioned that the firm's success could be traced to "the genuine Yankee 'knack' of doing things."[9] Banigan graduated to journeyman in that trade, working in Providence and sometimes in the bordering towns of Attleboro and South Attleboro, Massachusetts, all part of a dynamic and successful regional orbit of precious metal production with Rhode Island's capital city as the epicenter. Either during or after his apprenticeship he struck out on his own, creating a "corkscrewing" machine that intertwined gold with coral and shells for inlaid jewelry. Manufacturers widely used his

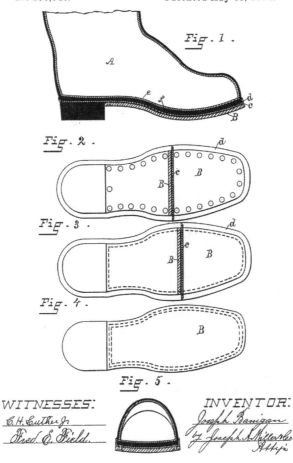

(No Model.)

J. BANIGAN.
RUBBER BOOT.

No. 298,546. Patented May 13, 1884.

Fig. 1.

Fig. 2.

Fig. 3.

Fig. 4.

Fig. 5.

WITNESSES:
C. H. Luther Jr.
Fred. E. Field.

INVENTOR:
Joseph Banigan
by Joseph A. Miller & Co.
Atty's

A patent for a highly specialized boot designed by Banigan for miners which combined leather and rubber soles fastened in a way so as to prevent puncture and leakage. (Courtesy Providence Public Library) invention to produce stock of these popular ornaments. Banigan might have made a good profit from his discovery and probably won some admirers and, perhaps investors for his future endeavors in the rubber industry.[10]

The Irish Fifth Ward

The Banigan family lived on Orange Street in Providence's Irish Fifth Ward, presently in the downtown area, with his father's brother-in-law, Joseph Finnegan, who had married the elder Banigan's sister. He was a fellow castaway and neighbor from Lisirrill on the Shirley Estate in County Monaghan who may have tended the small plot of land there while his relatives tried their luck in Scotland from 1845 to 1847.[11]

An average of nine people inhabited a couple of rooms in these stifling immigrant quarters in the Fifth Ward. Situated along the western banks of the Providence River, the immigrant neighborhood resembled a medieval castle with the harbor as a moat and a string of industrial buildings overlooking the water like a fortress. Behind the hulking but bustling edifices a warren of sheds, hovels, and tenements sheltered the area's seven thousand proletarian residents, probably not much better than County Monaghan's stone or sod homes (for the very poor) had protected its peasants. In fact, when a cholera epidemic spread through Providence in 1854, stoking fears of Famine-like afflictions, the Irish areas suffered disproportionately. Regardless of the circumstances, many managed to scrimp enough to send allowances home. One local employer paid fourteen dollars "to an Irishman who sends four pounds to his father." The residents of the Fifth Ward moved frequently, as many as three-fourths leaving the neighborhood between 1850 and 1860.[12]

An unusual description of Irish-Catholic life in Providence in this era survives in the valuable diary of a young priest from County Monaghan who raised funds for a Catholic university in Ireland. Father James Donnelly, born in Monaghan in 1823, visited the United States in the early 1850s and stayed in Providence, which hosted a considerable number of immigrants from that region. His local sojourn lasted from March through August, 1854. He observed "several cases of women driving buggies here. . . . Head–dress of woolen hosiery for all little girls." His first impression of the town was very positive: "Fine city[,] clean. No squalid poverty. Many public clocks. Number of churches, spires, etc. Yankeedom in earnest. Sermon of Bishop [probably Bishop Bernard O'Reilly, 1850–1856] Windbag. No argument at all. All words and no order. Still smart man. Great about rubrics."[13] A few days later the visiting priest cried on St. Patrick's Day listening to hymns that must have stirred his homesickness. He again criticized the sermon in the privacy of his diary. "Bishop preached at night. Poor affair on such an occasion."[14]

Father Donnelly celebrated mass at the Mercy Convent in Providence, close to the Banigan home. He received only fifty dollars at the first service but at the "High Mass[,] collection glorious"; he took in over $800. "Oh so good a people, especially the girls. God bless them." He also reveled in the local introduction to parishioners from his own diocese in County Monaghan who probably would never return to their homeland. Several of the children from the old country visited him while he was in the capital city.[15] He enjoyed the repasts but did not warm up to the church authorities here. "Remember seminary in Bishop's house, Providence. How learn here virtue or anything else? Hearing, seeing drinking etc. Ridiculous!" He witnessed an extravagant local Irish-Catholic funeral and observed a mother "keening" (public wailing for the deceased) in "old Irish fashion. Yankees listening. Remember large number of coaches, buggies etc. at funeral. This common practice. Will the Irish never have sense."[16] While in Providence, he wrote, "Sorry came here," and instead used the city as a transportation hub for train trips to other New England towns.[17]

Father Donnelly returned to Ireland in 1855 and eventually served as the Bishop of Clogher in County Monaghan (which included Banigan's township of Lisirrill) from 1864 until his death in 1893. In 1885 he received an unknown visitor. Banigan paid his respects and dues with a donation during a trip to his ancestral home: "Mr Jos Bannigan [sic] of Providence, R. Island visited, vastly rich, extremely generous— from near Ballitrain."[18] The Banigan family would doubtlessly have met Father Donnelly during his stay in Providence in 1854 and maybe Joseph remembered some long ago encounter.

Whatever the priest witnessed in Providence his hosts probably showed him the better parts of the town. Social and psychological problems in these urban, ethnic ghettoes triggered drinking and petty crime that, at times, spiraled into felonies, belittling the Irish community in the eyes of the wary Yankees for decades to come. The murder of a city watchman clubbed by one of four "midnight revelers" in the heart of Providence's Fifth Ward in May 1852, when Banigan still worked at the screw company, helped prepare the anti-immigrant sentiment of the upcoming Know Nothing period. The perpetrator, Charles Reynolds, apparently escaped to Europe, but others were arrested. The *New York Times* entitled the affair, "Irish Riot in Providence." The *Providence Journal* commented editorially: "This is the third person who has been attacked and beaten by gangs of drunken Irishmen within the last ten days."[19] The Board of Aldermen ordered watchmen to carry pistols, while a petition circulated to bring back capital punishment, which had ended in 1852, partially as a result of the backlash against the execution of John Gordon.[20]

A double standard, however, did exist. In 1853, more than a year after the murder of the Providence watchman, two groups of predominantly Yankee volunteer firefighters—allegedly intoxicated—fought each other for the right to extinguish several blazes in the capital city. Each incident led to escalating tensions. During this period, local bars and merchants often distributed free drinks to firemen to "fortify" them in their endeavors. The altercations reached a zenith in October when the volunteer companies turned their competitive wrath against each other. Ironically, Neal Dougherty, an Irish driver who was not formally a member of the Yankee crew, hit a rival. The other crew beat him to death yelling the usual "kill the d———d Irishman." There were no convictions, only a lingering battle of words in newspaper editorials. As a consequence that same year Providence established a police station in the Irish Fifth Ward and expanded night patrols, ostensibly to control immigrant troublemakers. Most of the officers, who held day jobs as well, felt the city's pressure to abandon moonlighting and work full time for the police department in order to professionalize the force. The Fire Department followed suit.[21]

Nationally, such incidents in the early 1850s helped form an unusual, anti-immigrant group known as the Know Nothing or American Party. The increased number of Irish-Catholic refugees only fueled the political outburst, like the thirty immigrants who became naturalized citizens at a publicized, official swearing-in ceremony in Providence in March 1851. The xenophobic Know Nothing movement, on the other hand, featured different but related issues like temperance, which attacked drunkenness in the Irish-Catholic community. In Rhode Island the social virus spread rapidly without the benefit of a toleration inoculation. Remnants of the Law and Order Party, aligned with conservative, rural Democrats, swept state elections in 1855 on a platform to extend the residency requirements for newcomers almost indefinitely. An incident that year resurrected the ghost of John Gordon. The Sisters of Mercy, recent arrivals in the state, established a convent and taught in parochial schools in Providence. Rumors surfaced, and appeared in local newspapers, that the nuns kept an unwilling novice against her will. Native-born citizens laid siege to the place, in the middle of the Irish Fifth Ward, but encountered a determined force of Irish Catholics. Bishop Bernard O'Reilly, sarcastically referred to as "Paddy the Priest," mobilized his flock to defend the institution. After some pushing and shoving, the crowd of a few thousand dispersed with little damage and few injuries.[22]

The Know Nothing phenomenon, a kind of precursor to the Ku Klux Klan, ended almost as quickly as it started nationally but lingered in areas with Irish-Catholic communities. One long-term effect of the movement was to equate Protestantism with Americanism. Father Donnelly, in Boston at the time of the election, wrote that "'Know-nothings' gained everything by immense majority. Awful times at hand I fear." When an immigrant reported a crime to the Providence police in 1859, a local newspaper, reflecting the temper of the times, mocked the event: "Yesterday afternoon, Patrick O' Some-thing-or-other, we don't remember what, and it's certainly of no consequence, called at the Police office to make complaint against another Patrick whose name was Michael, saying that the last mentioned individual had knocked him down." That same year Thomas Davis, the Rhode Island Congressman, lost an election when anonymous handbills urged voters to select another Protestant candidate "INSTEAD OF AN IRISHMAN."[23]

The numbing cultural insensitivity only hardened feelings between old-time residents and ethnic newcomers. The bitterness and animosity limited employment for the Irish denizens of the distressed Fifth Ward. Still, many obtained menial jobs—especially children—a short walk away. Prominent among the neighborhood establishments was a factory of the infamous Sprague cotton empire, now under the control of the son of the martyred Amasa. The American Screw Company also had one of its buildings here where Banigan toiled as a youngster. The Providence jewelry industry moved into the district around 1850, and onto Orange Street itself, according to Rhode Island historian Thomas Bicknell. The many shops, now just a few doors away, may have offered convenient work to the increasingly talented Banigan. Bicknell stated that: "Most of the leading jewelers have come up to leadership from the benches, where they learned the details of the factory. Brain work and skill are essentials to success."[24] More important for Banigan's future, the waterfront hosted the Providence Rubber (Shoe) Company at the corner of Dorrance and Clifford Streets, again just a few blocks from his home, where he actually started his career in rubber for a short period around 1860.

The Rubber Industry

Rubber was often called India-rubber in the nineteenth century, after South American natives discovered the substance and used it to make bouncing balls for sports, as noted during the second voyage of Columbus. The appellation "rubber" came from the famous English chemist Joseph Priestley, who originally employed the little understood substance to "rub" out stray markings from his pencil just before the American Revolution. (Banigan would also produce pencil erasers for a short period at the beginning of his own career in Woonsocket and actually held a patent.) Quality rubber trees grew almost exclusively along the Amazon River in Brazil in ideal rain forest conditions. Native gatherers tapped the trees in a way similar to the harvesting of maple syrup. The latex secretion (not the inner sap), actually healed any incisions to the tree. Pale white in color, the "jungle milk" became black when heated and dyed. Most of the world's supply came from an area known as Para at the headwaters of the Amazon. Initially Brazilian entrepreneurs sold primitive footwear to New England merchants, including crudely crafted waterproof shoes they wore themselves. Massachusetts merchants and shippers, especially in Salem, monopolized the beginnings of this trade in the 1820s and 1830s, but eventually imported the gum directly to New England and applied the substance to boots. Some reversed course and established shoe manufacturing virtually in the Brazilian forest at Para, which eventually became a thriving city, by shipping precision-made lasts to properly shape the footwear there.[25]

The chemical harnessing of rubber was a complex operation, and the legal battles emanating from the question of who actually created that formula proved just as consuming. Inventors sought a way to stabilize "the treacherous, intractable product" to avoid its melting in the heat or cracking in the cold.[26] A couple of entrepreneurs who would show up in Providence and play a role in Banigan's empire experimented with rubber coated fabrics in Roxbury, Massachusetts, in the early 1830s. Edwin Chaffee would invent important machinery for the manufacture of rubber while John Haskins would enter a partnership with Banigan during the Civil War, making rubber stoppers for bottles and containers. Meanwhile, two other cohorts, Nathaniel Hayward and the famous Charles Goodyear, subjected raw rubber to various experiments. Goodyear mixed it with sulfur and lead. In 1839 he carelessly baked it on his home stove, inadvertently discovering the secret to controlling the substance through heat. He called the process vulcanization after Vulcan, the Roman god of fire. Although he discovered the way to harness rubber, he spent many more years perfecting the curing process.[27]

As soon as Goodyear patented his experiment in 1844, challengers sued. As the case, and about sixty related ones, wound their way through a labyrinth of judicial tribunals, Goodyear sold manufacturing licenses in return for moderate royalties. Many of the patentees banded together to collectively fight infringements. The indefatigable Daniel Webster defended Goodyear's claim in court in "The Great India Rubber Case" of 1852 for a fee of $25,000, said to be the largest legal compensation in American history at the time. Goodyear finally had his patent for vulcanization protected by a major legal decision in his favor. Others tried to circumvent Goodyear's stranglehold on the process, but his discovery became judicially ironclad until the patent's expiration in 1865.[28]

Rhode Island Rubber Footwear

In the 1840s–1850s according to the authoritative *Boot and Shoe Recorder,* "the bulk of the rubber shoes manufactured in the United States were universally known as 'Providence shoes.'" Some of the state's leading historians fail to mention this pathbreaking and important industry, although the footwear remained fairly primitive until the vulcanization breakthrough. The Providence Rubber Company was one of four major footwear producers in the city before the Civil War and, in that era, employed some of the greatest names in the industry. George Bourn, father of future Rhode Island Governor Augustus Bourn (1883–1885), opened his establishment in 1840. A decade later Edwin Chaffee, a partner and confidant of Charles Goodyear, joined Bourn's firm. Chaffee, a master craftsman and inventor, revolutionized the industry as much as Goodyear but with his own patented mammoth machinery. The Bourn factory legally produced footwear independent of the Goodyear process, through a slightly different curing process. However, the Providence Rubber Company and two of the other leading city manufacturers, eventually became defendants in infringement suits after both the elder Bourn and Charles Goodyear died in 1859 and 1860, respectively. Suits and countersuits swirled through the rubber industry after the inventor's death. The Providence action went through Rhode Island U.S. District Court, ending in a favorable judgment for the local firm, only to be overturned by the United States Supreme Court on appeal. The Court assessed Bourn's company more than $300,000 in damages in 1867, two years after the Goodyear patent expired. Bourn's son Augustus changed the firm's name and built the National Rubber Company in Bristol, Rhode Island, in 1864. The enterprise diversified, as it already had started to do under the old regime in Providence, producing rubber blankets, tents, and boots during the Civil War, as well as clothing, druggist supplies, and other products. Not long after the Supreme Court decision Bourn folded the capital city enterprise and merged what was left of the old business into his Bristol facilities.[29]

At the beginning of the Civil War Joseph Banigan worked at Bourn's Providence Rubber Company, where he met his future father-in-law, John Holt (1824–1896).[30] Holt grew up in Scotland and worked in a textile factory in Glasgow at the age of nine (not too far from Dundee where the Banigans spent two years), before the family moved to Manchester, England, where he continued in the same line of employment. He emigrated to the United States in 1851 and joined the Providence Rubber Company. He became a department head and later superintended boot and shoe operations in the Bristol facility. He would serve in the same capacity with Banigan in his forthcoming venture at the Woonsocket Rubber Company. Earlier in his life he converted to Catholicism, although it is not clear whether that occurred before or after he left Great Britain. Nonetheless, the status of religious turncoat must have been an albatross for this English immigrant in Yankee Rhode Island. He lived at Manchester Place on the outskirts of the Irish Fifth Ward in Providence, a half dozen blocks from Banigan. (The street name, Manchester Place, may have had something to do with Holt's emigration from that town, or maybe a small colony of émigrés lived there. Manchester, England, was also home to Charles Macintosh, who experimented with rubber varnish for the waterproofing of clothing, thus the label, a macintosh.) Holt testified for his employer in the Goodyear suit, discussing his duties as prosecutors established workplace practices and processes to prove a patent violation. He apparently possessed a wide set of managerial, chemical, and production skills, assets that paralleled Banigan's and eventually dovetailed into a brilliant partnership.[31]

Probably around 1860 Banigan finished his apprenticeship in the jewelry trade and served that fateful short stint at the Providence Rubber Company where he met John Holt. Michael Joseph Banigan married Holt's daughter, Margaret, on 26 December 1860. He was twenty-one and she was seventeen. Bishop Francis P. McFarland conducted the ceremony, another indication that Banigan stood out from the crowd as well as the congregation. Banigan was already challenging fellow parishioners to follow his lead and donate more generously to St. Joseph's Church in the Fox Point section of Providence. Irish immigrants founded that parish in 1851 in an area almost as heavily Irish as the nearby Fifth Ward. Banigan's funeral was held here in 1898. By 1859 the family, along with the Finnegans, their in-laws from County Monaghan, moved to Winslow Place at the edge of the Fifth Ward across the street from Saint Xavier's, site of the earlier Know Nothing riot. Bernard Banigan, Joseph's father, was listed as a laborer until his death in 1867, the same as on his emigration papers in 1847. Laborer remained a common appellation for unskilled workers, usually of Irish origin, at that time. Finnegan worked as a tailor just a few blocks away on Eddy Street. Those who toiled in the service industry, providing merchandise and assistance to fellow ethnics—their own kind—seemed to prosper.[32]

Soon afterward Banigan became a partner with John Haskins in making rubber plugs for bottles and containers in the Boston area. Haskins was a founder of the pioneering Roxbury India Rubber Company and, as a confidant of Goodyear, held one of the earliest licensing patents from the great inventor, who sometimes conducted his latest experiments there. Edwin Chaffee, the master mechanic, was also a partner in the Boston enterprise. In its obituary of Banigan in 1898, the *India Rubber World* wrote that the association with Haskins, and presumably a constellation of other rubber virtuosos "was destined to change the whole tenor of the young jeweler's life and place him in the front rank among rubber-manufacturers."[33] An amateur historian in Woonsocket concluded that Banigan mastered Haskins' abilities and grafted them "to the other skills he had acquired from his years in a machine shop and in the jewelry trade. He understood rubber and he understood machines."[34]

During the Civil War Haskins and Banigan established the Goodyear India Rubber Bottle Stopper Company, in reality a one-room laboratory with a solitary rubber grinder on Eustis Street in Roxbury, Massachusetts. Banigan made a modest salary of ten dollars a week but enjoyed the title of superintendent of the little empire. With the energy that characterized his career, Banigan continued to add machinery and boost production, so much so, that he and Haskins single-handedly glutted the market with rubber stoppers in a few months! He then fanned out across the region seeking orders for related rubber goods, which he produced at the factory. According to the *India Rubber World*, he "acted as salesman, superintendent, and factory help," earning a 5 percent commission of $1,500 the first year. Because of his skills Banigan could compress several layers of management functions into his own personal portfolio. He embraced the era's fascination with being a jack-of-all-trades, the commercial everyman who knew the nooks and crannies of an industry and never flinched regardless how Promethean or menial the task at hand.[35]

This peripatetic energy helped finance a new plant in Jamaica Plain, Massachusetts, under Banigan's direction. During his tenure there, he apparently rose early to conduct experiments in a tiny facility at home. Not unlike his invention of the corkscrew jewelry process, his efforts paid off when he discovered a way to cure rubber rollers without sulfur. He sold the rights to the process for $3,000. Part of the

Banigan legend was that this chemical formula bypassed Goodyear's vulcanization method, although any such notoriety would have been fleeting because the master's patent had already expired in 1865.[36]

Three of Banigan's four children were born in the Boston area: Mary in 1861 when her father was still listed as a jeweler, John in 1863, and William in 1864 when the birth certificate described Banigan's occupation as a rubber manufacturer. The fourth child, Alice, named after his mother, was born in 1866 in Woonsocket, Rhode Island.[37] Banigan also permanently dropped his Christian name, Michael, and used his middle name exclusively, a common Irish practice.

The Civil War Era

While Banigan immersed himself in the intricacies of vulcanization and the complexities of running a business, the United States fought a civil war over the issue of slavery and states rights. Although Banigan stayed put at his shop in Jamaica Plain, many of his compatriots, especially in New England and even the Fifth Ward, joined the Northern force. While some Irish-American soldiers fought for the noble cause of abolition and the maintenance of the union, many enlisted to earn liberal bonuses unmatched in industry, despite the battlefield danger. No one questioned the bravery of these volunteers, who often banded together in predominantly ethnic regiments and fought ferociously. Some Irish nationalist leaders viewed the conflict as a way for rank and file soldiers to gain military experience for an eventual foray back to Ireland to foment a revolution.[38]

As in other states, the Rhode Island Irish tried to form a single Civil War unit, the Sarsfield Guards, to house their numbers. They may have been encouraged to do so by Governor William Sprague, the son of the murdered Amasa. Several local newspaper articles noted the attempt to forge an exclusively Irish Third Regiment with the establishment of at least four companies of volunteers. The effort failed, apparently from a lack of Irish officers, internal fighting, and opposition among the local Yankee military hierarchy, with the Dorr Rebellion in mind, against forming such a segregated force. However, the unit did march on St. Patrick's Day in Providence in 1862. More than 1,700 Irish-born soldiers eventually composed one-fourth of the Third Regiment's huge roster. They suffered 196 casualties. A local priest who

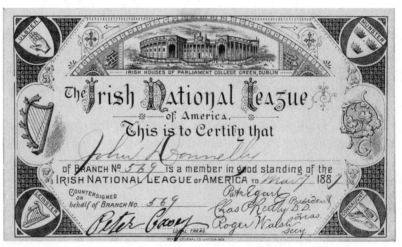

HEAD QUARTERS,

EMMETT GUARDS, CO. D.,
5th Battalion Infantry.

Providence, *Nov 2?* 18 7 5

Mr. *Horse Shoer Union* to Co. D. Dr. $ 2.00

On account of *Hall Rent & Sat.*

Received Payment, *James E Curren* TREASURER.

A receipt from the Emmet Guards for rental of the battalion's hall to the Horse Shoers Union, a mostly Irish labor group in 1875. The formation of an Irish company in the state militia took some of the sting out the inability to form an exclusive ethnic force from Rhode Island in the Civil War. (Author's Collection)

ULSTER   MUNSTER

IRISH HOUSES OF PARLIAMENT COLLEGE GREEN, DUBLIN

The Irish National League
of America.

This is to Certify that

John Donnelly

of BRANCH № *569* is a member in good standing of the IRISH NATIONAL LEAGUE of AMERICA to *May 7* 188 7

COUNTERSIGNED
ON
behalf of BRANCH No. *569*

LEINSTER   CONNAUGHT

Patr Egan President
Chas Reilly D.D.
Roger Walsh Treas.
Peter Casey Secy.
LOCAL TREAS.

STATE JOURNAL CO. LINCOLN, NEB.

John Donnelly's membership certificate in the Irish National League in Pawtucket in 1887. Innumerable patriotic organizations permeated Irish-American communities throughout the nineteenth and early twentieth centuries. (Author's Collection) served briefly with the force wrote to inform Bishop McFarland that his participation would help "a good deal in annihilating the Protestant prejudice in Providence."[39]

Overall, Irish volunteers served in all twelve Rhode Island regiments. Almost 6,000 of the 23,547 soldiers from Rhode Island were first- or second-generation immigrants, the lion's share presumably from Ireland. Several local Irish priests served as chaplains but there was also an antiwar undercurrent within clerical ranks. Father Michael McCabe, later a confidant to Banigan in Woonsocket and vicar general of the church in Rhode Island, allegedly advised Irish Catholics not to enlist. Newspaper editorials here and abroad commented on the exploitation of both Irish and Black regiments as cannon fodder, suffering abnormally high casualty rates during the war. When the South surrendered, Congress created the Medal of Honor to highlight exceptional courage. Two Irish Americans from Rhode Island, John Corcoran and James Welsh, received the award.[40]

A move to provide the right to vote for all Civil War soldiers began, at the start of the war in 1861, to spur even greater enlistments by the Irish and other "adopted citizens" in Rhode Island. Although the measure passed the General Assembly, with some discriminatory public remarks in the Senate, voters rejected a constitutional amendment in 1863, in the middle of the conflict. One flyer caustically asked about the indirect impact the rejection would have upon Irish-American soldiers: "Should this act be defeated, how think you the men of the glorious Third R. I. V., who are bleeding in the trenches before Charleston, will feel?" Voters turned down a similar bill in 1871, vocally questioning the loyalty of foreigners and Catholics despite their unquestioned valor on the battlefield. Altogether Civil War veterans witnessed the measure fail three times. They had fought for many different reasons, to be sure, but found themselves still blacklisted when it came to voting privileges. Too many opposing interests, mixed with historical prejudice, prevented the empowerment of Irish Americans. Yet the Irish became active in some Grand Army of the Republic veterans' chapters. In Woonsocket the Ancient Order of Hibernians invited Irish-born General James Shields, the only person to serve as a U.S. Senator from three states, to speak. He stayed at John Holt's home. When the voting amendment finally passed in 1886, more than twenty years after the southern insurrection, some soldiers had already left for a celestial muster. By that time the Irish had formed the Emmet Guards, Company D, 5th Battalion of Infantry in the state militia as a kind of ethnic showpiece. According to Rhode Island historian Charles Carroll, author of a multivolume history of the state, the rebellious nature of the Irish in Ireland found fertile ground in American politics. They "longed to participate," he wrote, "but crashed against the shoals of Yankee discrimination." Furthermore, the Irish flocked to the local and discredited Democratic Party. The Yankees, by and large, ignored and forgot their sacrifice. The Irish frequently abreacted by redirecting any disappointment into local nationalistic and Fenian causes.[41] During the Reconstruction era a strange echo of the Irish question in the Dorr War reverberated in a national controversy. In the congressional debate to ratify the Fifteenth Amendment to the United States Constitution in 1869, Henry Bowen Anthony, *Providence Journal* editor and now a Senator from Rhode Island, continued his anti-Irish-Catholic efforts. He successfully argued against including the word "nativity" among the Amendment's provisions to prevent voting discrimination. Supporters required Anthony's backing and the state's vote in order to counterbalance southern opposition. The attempt to include the Irish in Rhode Island and the Chinese in California under the law's protected categories failed in order to gain Anthony's support for the original provisions to enfranchise freed slaves. Ironically, local African Americans secured the right to vote because of the Dorr War in 1842, so the passage of the Fifteenth Amendment in 1870, stripped of any ethnic component, had no impact on the state.[42]

The Civil War had drained manpower from local factories. Northern troops from New England had consisted mainly of Yankees, Irish, and free African Americans. Smaller groups from the ranks of the old immigrants also joined: especially Germans and Scandinavians (mostly Swedish in Rhode Island). On the home front, where employers searched desperately for new hires, a mighty recruitment effort centered on Quebec, where an economic pinch in the dominant agricultural sector there ignited one of New England's greatest waves of immigration. With direct railroad service from one region to the other, thousands of French-Canadian farmers answered the call from New England's manufacturing plants. Heads of households eventually brought their entire families as economic units.[43]

French Canadians had no initial desire to stay in the United States, embracing a culture of *survivance,* a trinity composed of their Catholic faith, French language, and loyalty to French Canada. They came and went like snowbirds between the two worlds of farming and textile production. Eventually more settled south of the border in New England and by 1900 had wrested demographic control from the Irish in Banigan's Woonsocket. Unlike their Irish-Catholic brethren who controlled the Church and usually disliked their Yankee employers, French Canadians remained less combative toward mill owners (some of whom hailed from France) and even envisioned a shared community of interest between toilers and titans—a joy to the local Republican Party. The GOP further applauded the struggle between French Canadians and the Irish for control of the Catholic Church, which helped prevent a political alliance between the two immigrant groups. Although newcomers from Quebec dispersed among the economic valleys of the state, they still congregated in textile mill colonies. During times of slack employment, they could travel by train back across the border to work temporarily again on family farms, an option unavailable to almost all other immigrants. The alleged docile nature of these mid-century arrivals, perhaps reinforced by the belief that their stay was temporary, provided little protection from the usual nativist demand that all immigrants assimilate quickly. But the French language, transformed into grammatically impaired, accented English, provided a wealth of jokes and ridicule that still make the contemporary circuit today. And it troubled some hard-line Yankees that French Canadians practiced a more intense version of Catholicism than the Irish, even though the quarrel between the two immigrant groups fractured any Catholic unity and contributed to local conservative dominance. The initial lack of interest in citizenship by Quebec natives weakened them in the political arena but also made them less of a target for the more active, higher profile Irish Catholics.[44]

Because of the intense discrimination against local Irish Americans, whose economic opportunities remained almost as limited as those of their counterparts from Quebec, they remained close to the bottom of society's pecking order, where they fiercely competed with French Canadians and African Americans for unskilled jobs. In the 1860s, local newspapers often carried reports of fistfights and brawls between Irish immigrants and blacks, fueled by a mixture of mutual racism and economic struggle. In 1867 African and Irish Americans fought at the Rocky Point amusement park, and in Providence a gang of "Irish 'skinny' boys" attacked two black men. Ironically, the Dorr War in 1842 had propelled African Americans to a significant political lead in the contest between the two groups. Blacks gained the right to vote when conservatives reached out to them to counterbalance the almost imaginary Irish-American participation in the Dorr upheaval. The immigrants, with much less seniority in the country than the scions of early slaves, faced many obstacles to cast a ballot. The hostility between the two increased as the number of Irish newcomers proliferated in the years before the Civil War. The first state census in 1865 noted that the Irish in Rhode Island numbered 27,000 in a population of almost 185,000. The combination of first- and second-generation Irish, however, totaled 3 of every 8 citizens. African American numbers, on the other hand, remained fairly small at 1,700 in Providence and 4,000 statewide.[45]

Here and there a brave voice on either side of this ethnic and racial divide spoke out against the oppression facing the other. For example, a black lawyer, William H. Bell, attended an Irish nationalist gathering to establish a branch of the Irish Land League in Cranston in 1880. He took the floor to compare the oppression of both races in a "stirring" speech. Earlier, Frederick Douglas, the black civil rights crusader, supported the decision of African Americans to achieve the right to vote locally when he visited Providence at the time of the Dorr War, and on that occasion endorsed the uncomfortable alliance of blacks with Law and Order conservatives. But Douglas also defended civil rights for the Irish on several trips to Ireland where he spoke on behalf of his own race. "Never did human faces tell a sadder tale," Douglas observed at a large meeting in Cork, Ireland; "these people lacked only a black skin and wooly hair to complete their likeness to the plantation negro. The open, uneducated mouth—the long gaunt arm—the badly formed foot and ankle—the shuffling gait . . . all reminded me of the plantation, and my own cruelly abused people."[46]

The great pan-African scholar and activist W. E. B. DuBois wrote in his autobiography about the surprising prejudice he witnessed growing up in Massachusetts in the Gilded Age: "The racial angle was more clearly defined against the Irish," he remembered, "than against me." These voices in the wilderness remained just that and, for all we know, may actually have embarrassed some Irish immigrants who felt greater alienation if not humiliation by having the descendants of slaves champion them. Some scholars maintain that Irish immigrants ran a blistering gauntlet before they became officially white themselves.[47]

On another front the Irish were often their own worst enemies, eschewing any meaningful coalition with other Catholic immigrants, particularly French Canadians. The Church in the state always possessed an Irish hierarchy and, to this day, with one exception, every Bishop has been Irish, causing consternation among other groups. Although Italian newcomers did not arrive in great numbers until the end of the Gilded Age, eventually equalling the Irish in the 2000 census, the Irish tangled with Italian nationalism and anticlericalism as early as 1860. A group of vendors on Federal Hill, an Irish stronghold before the Italians later inhabited that area, hung small banners from their carts saluting the father of Italian unification, Giuseppe Garibaldi, who had forcibly annexed the Vatican. "The Irish inhabitants of the neighborhood," according to one newspaper, "resented this as an insult to the Pope." Friction at the highest levels of the Catholic Church would contribute to conflicts and antagonisms elsewhere.[48]

When Joseph Banigan relocated back to Rhode Island after the Civil War, he probably could not realize the impact that the intersection of enterprise and ethnicity would have on his employees and his own career.[49] Banigan finished the construction of the modest factory at Jamaica Plain in 1866 but already had made plans to strike out on his own. He would move to Woonsocket, Rhode Island, to establish and propel a remarkable establishment that brought him untold honors and notoriety. A quarter of a century later, in 1894, he stood at the top of the world of rubber footwear as president of the fledgling cartel, the United States Rubber Company.

THE WOONSOCKET RUBBER COMPANY

. . . .

> *"When the rubber comes out from the oven . . . a marked*
> *change has taken place in its constitution, it is no longer*
> *dead, but possesses a marked vitality."*
> —Woonsocket Patriot, *20 April 1877*

> *"Mr Banigan has the full confidence of stockholders, is a*
> *shrewd capable man & has made considerable money."*
> *—Anonymous credit investigator for*
> *R. G. Dun & Company, 15 February 1882*

When Joseph Banigan turned twenty-five in 1864 he already had abandoned whatever it was in the Irish psyche and emigrant experience that kept so many of his compatriots behind. His youth and early work experience seemed to have opened new vistas for him. Banigan's inventive knack must have caught the attention of others who saw beyond the ethnic and religious blinders that too often dressed the Irish in stereotypes. Banigan's vision and enthusiasm for his work, which, in other Hibernian immigrants, might have been dismissed as pure blarney, coincided with his ability to get things done.

When, in 1867, a young Irish priest, the Rev. Francis P. Lenihan, died prematurely in Woonsocket at age thirty-three not long after Banigan set up shop there, the local newspaper wrote that "he did not believe in keeping the Irish isolated, which is to make them clannish; he wished them to mingle with Americans, to imbibe American views."[1] Banigan appeared to have embraced this philosophy. He eluded the gravitational pull of self-pity that came from deep in the earth of Lisirrill in County Monaghan, or any other geographical spot in Ireland for that matter. Although he escaped the visible pain that accompanied most expatriated Irish refugees, in reality he only side-stepped it. He avoided Civil War battlefields, life imprisonment in a narrow ethnic neighborhood, and false relief at a local tap. But he was no turncoat. He battled for his people against Yankee prejudice in the most inhospitable territory: the world of business and finance. And although his endeavors brought him wealth and treasure beyond his dreams, he never forgot his roots and always actively sought to improve the conditions and opportunities of his poorer comrades even when they compounded problems by their own uncivil behavior. But everything was on his terms, and he fought his own kind, as in the rubber strikes of 1885, with even greater ardor than his engagements with the state's Yankee ruling class.

At the age of twenty-five, Banigan was already more American than Irish and unharnessed by the deference and bitterness that fettered the abilities of so many other immigrants. He prepared to leave his salaried position in Jamaica Plain with John Haskins, one of the great pioneers in the rubber industry, and strike out on his own. He finished his work at the bottle stopper plant by overseeing the construction of a new factory there. At the same time he consummated a deal that would bring him to Woonsocket to run his own enterprise and pursue the quest of being an independent proprietor, an urge that had animated him since his teenage years.

The Bailey Wringing Machine Company, on Social and Clinton Streets in Woonsocket, made primitive mechanisms featuring novel rubber rollers to dry or "wring" clothes by squeezing the water from the freshened laundry between two coated cylinders operated manually by a crank. A partnership purchased the rubber components elsewhere but assembled the devices in Rhode Island's northernmost city, next to the Massachusetts border. Local "venture capitalists" took an interest in the operation and helped raise $250,000, selling shares almost exclusively in the Woonsocket area in $100 denominations. By 1876, sixty employees were producing 75,000 machines annually at a rate of over 1,000 wringers per worker. The firm used about 600,000 board feet of pine, birch, beech, and maple yearly. Demand still outstripped supply. With a monopoly of the patents for the washers, the owners controlled the national market.[2]

Woonsocket Rubber Company

After the wringing company opened in 1864, one of the new partners began making the crucial rubber covers for the rollers in an independent effort that soon supplied the required product to the parent company. Ironically, Banigan's father-in-law, John Holt, had held
the patent for that rubber roller device since 1865 while an employee
at the Bourn enterprise.[3] The local investors, Lyman and Simeon
Cook, "found" Banigan "in a small rubber factory near Boston" and
asked him to run the fledgling Woonsocket Rubber Company, originally conceived as a supplier to the wringing business but soon out-
stripping it in products and output. He apparently finished his work
with John Haskins over the next two years while helping matters along
at the same time in Woonsocket before assuming the position in 1866.
He lived in Woonsocket until about 1874 when he moved to Providence permanently to take advantage of the frequent and direct boat
and train service for business purposes. When he needed to stay in
Woonsocket he probably spent the night with his father-in-law.[4]

Initially, the Woonsocket operation was almost as primitive as the
original Haskins endeavor. Banigan entered a partnership with Lyman
Cook, a machinist, self-made entrepreneur, and Republican activist.
They raised $10,000, probably on the reputation of Cook, the home-
town investor with many financial and institutional connections, who
assumed the presidency of the company until Banigan bought him
out in 1882. Initially, they rented an old planing mill in town and bur-
rowed through layers of leftover wood shavings to anchor their meager hardware: two grinders, a calender (a rolling machine to press and
smooth the rubber), and a fifteen-horsepower engine that required
constant attention. At first the company produced only rubber rollers
and blankets, but Banigan continued what was to be a lifelong practice of upgrading machinery and enlarging facilities whenever possible. The hard-bitten credit operatives of R. G. Dun and Company
later marveled at Banigan's ability to pay large dividends ranging from
25 to 35 percent despite ongoing expansion.[5] On 7 June 1867—Banigan's twenty-eighth birthday—the partners incorporated the Woonsocket Rubber Company capitalized at $100,000, a long way from the
family's semiannual rent of 3 shillings on the Shirley Estate. Banigan

The Woonsocket Rubber Company about 1870. Recently the building was totally renovated. (Courtesy Harris Public Library, Woonsocket)

borrowed $30,000 on his own merits. A month earlier he had taken the oath to become a naturalized United States citizen. Joseph Finnegan was a witness. Curiously, Banigan and his two sponsors all wrote that he arrived in Rhode Island in 1854 rather than 1847; perhaps there was some requirement that made a later arrival more advantageous, either personally or before the law. In his affidavit Banigan stated he was born "around the year 1840."[6]

Soon after, in 1868, the firm started producing boots and shoes almost exclusively and left the business of crafting rubber rollers altogether, although Banigan eventually controlled the Bailey Wringing Company and would continue to dabble in the manufacture of other auxiliary products. The *India Rubber World,* in its Banigan obituary in 1898, wrote about the Woonsocket years: "Mr. Banigan was the practical rubber-worker. All the machinery from the start was set under his direction, and in the early days he superintended the factory and sold all the goods."[7] Furthermore, Banigan held at least four patents from the time of his employment in Woonsocket, including the very resourceful process to combine rubber and leather boots for miners in 1884.[8] Charles Carroll, a sympathetic Catholic who published a four-volume history of the state in 1932, credited Banigan with possessing "a Midas touch."[9] On the other hand, a more contemporary local "Cyclopedia" of Rhode Island notables in 1881, which did not include Banigan and listed only a handful of Irish Catholics, mostly clerics, lionized Lyman Cook, the principal investor and president of the Woonsocket Rubber Company. The editors praised Cook's considerable and deserved real estate, banking, and manufacturing expertise. Ironically the language and sympathy reflected the sentiments so evident in the anti-Irish letter by James Wilkinson in 1901, previously cited. The writers commended Cook for "giving employment to many, and contributing materially to the growth of this and other localities." Banigan still had to wait his turn, although within a year of running the fledgling enterprise he doubled capacity, constructed specialized buildings for vulcanization, and installed a one-hundred-horsepower engine. Employment jumped from fifty to one hundred as the Irish influx began. Banigan altered the lease and bought the property outright, the operation was assessed at $41,000 in 1868.[10]

Woonsocket 1865–1875

The town of Woonsocket was carved out of the town of Cumberland in 1867 and was later incorporated on 13 June 1888. Woonsocket had a population of about 5,375 after the Civil War with 3,500 factory jobs that drew in "hands" from surrounding communities. Labor agitation stirred the operatives, especially about the ten-hour day issue, for a generation.[11] In 1875 the population more than doubled to over 13,000, although 3,000 came from the annexation of its western part from the town of Smithfield by an act of the General Assembly in 1871. More important, Woonsocket had the highest rate of foreign-stock inhabitants in the state at a whopping 72 percent of the population. The Irish and French Canadians each made up about one-third of the new entity's residents. American-born citizens constituted less than one-third (and some of them were children of earlier immigrants). Those of English, Scotch, and Welsh origin numbered just over 700.[12] A local branch of the American Protective Association, dedicated to curtailing immigration, held an organizational meeting there in January 1868, but their influence proved stillborn as a result of the sheer number of newcomers. When several letters to the editor argued caustically about the genesis of riots between Protestant Orangemen and Irish Catholics in New York City in 1871, the *Woonsocket Patriot* refused to print correspondence about the issue, arguing that the newspaper was "not a medium for the discussion of theology." No one wanted ethnic and religious strife in such a polyglot city.[13]

Beginning in 1810, Woonsocket hosted a growing textile industry that specialized in cottons. The Blackstone River, which gave its name to the surrounding valley, provided energy for a serpentine necklace of factories reaching from Worcester to Pawtucket, the site of Slater's Mill. By the time Banigan arrived, the area featured several major textile manufacturers and allied businesses that marked Woonsocket as a place of commercial importance. By 1878, fourteen cotton mills and four woolen plants employed almost 3,400 workers. The Providence and Worcester Railroad (1847) transported the fruits of the loom as effortlessly as the Blackstone River churned the water wheels to make the goods until steam engines supplanted them. The P&W, as it was popularly known, continually made improvements to rails and service, especially double-tracking the most patronized passenger and freight routes—an important requirement for the ever-increasing shipments of goods to and from the city. Banigan's deliveries of raw rubber would later tax even the improved rail facilities. Primitive omnibuses and stagecoaches, however, continued to ply the roads between adjacent villages.[14] Edward Harris (1801–1872) was the proprietor of the largest woolen mill and provided an example of *noblesse oblige*. He set a standard of philanthropy that earmarked him as an unusual Yankee entrepreneur, donating $500 for Famine relief in 1847. On the political scene, he ran for governor several times as a staunch abolitionist and hosted Abraham Lincoln in 1860, also setting him apart from most of his peers, who feared a break with the cotton South that provided so many northern mills with raw material and a market for inexpensive slave cloth called kersey. Although Harris and Banigan only resided in Woonsocket together for a few years before the former's death, the long shadow of Harris' benefactions may have encouraged similar predilections already present in the Catholic rubber maker.[15]

The Irish in Woonsocket

The Irish appeared in the region in the 1820s, toiling, like Michael Reddy, as laborers on the Blackstone Canal from Providence to Woonsocket and Worcester. Many settled along the way and became factory operatives themselves. During the Civil War thousands of French Canadians from the Quebec area served as substitute workers for enlisted soldiers throughout the mills of New England, fabricating equipment and textile products for the Union. Although most of these second-wave immigrants had no intention of settling here permanently, over time they remained and gave a new accent and culture to many manufacturing villages as well as reinforcing the growing number of Catholics in the region. Banigan arrived just as the Irish and French-Canadian populations perfectly balanced one another in Woonsocket. Although demographics soon favored the settlers from Quebec, Banigan's unofficial policy of partiality toward Irish employees triggered an invasion of his compatriots to the Woonsocket Rubber Company as the concern expanded. Generally the early Hibernian entry, and the size and stability of the Irish workforce, resulted in real estate, voting priv-

ERIN, THE LAND OF OUR BIRTH.

ST. PATRICK'S DAY!!

OUR NATION'S JUBILEE WILL BE CELEBRATED BY A

GRAND BALL

— AT THE —

UNION HOUSE HALL, BLACKSTONE,

ON WEDNESDAY EVENING, MARCH 16, 1859.

A cordial invitation is extended to all, especially to the sons and daughters of *Erin*, who love her though in sorrow, that on this festive night, we may recall the many endearing associations, and rekindle the fires of patriotic love, which we owe to the Land of our Fathers.

GENERAL COMMITTEE.

Richard Mason,	Blackstone.	Ed. Cunningham,	Blackstone.	Samuel Almy,	Woonsocket.	Samuel Davis,	Slatersville.	John Kelley,	Pascoag.
Francis Kelley,	"	James Cakill,	"	Thomas Takel,	"	William Jackson,	"	John Quirk,	Harrisville.
Dennis McMullon,	"	Edward McMahon,	"	Henry Kelley,	"	Michael Gafney,	Uxbridge.	Patrick Gilligan,	Glendale.
Patrick Cook,	"	Michael McKone,	"	George Miller,	"	James McGalby,	"	Richard McNelly,	Providence.
William Kelley,	"	John Dolan,	"	John Geoghegan,	Millville.	William Maguire,	"	John Fowler,	"
Mack Mulgrew,	"	Lawrence Danbey,	"	George Bryan,	"	Patrick Cunningham,	"	Michael Hayden,	"
Charles Crispen,	"	Timothy Lawrence,	Woonsocket.	James McGinniss,	"	Patrick Houghton,	E. Douglas.	Peter Goodfellow,	"
Patrick Haughes,	"	William Byrne,	"	Francis Coughlin,	"	John Conner,	"	Francis E. Kelley,	"
Patrick Hopkins,	"	William Williams,	"	Samuel Farris,	Slatersville.	Christopher Kirk,	Pascoag.		
John Quinlin,	"	John Masterson,	"	John Conlin,	"	John Conner,	"		

FLOOR MANAGERS.

JAMES E. SLATER, | WILLIAM KELLEY, | PATRICK COOK, | HENRY KELLEY.

MUSIC BY - - - RICHARDSON'S BAND.

Tickets for Dancing, - - - 50 Cents.

Supper Furnished to Order.   Carriages Furnished Free for Ladies.

ENOS HAYWARD, Proprietor.

From the Patriot Steam Printing Establishment, Main Street, Opposite the Depot, Woonsocket.

Joseph Banigan originally built his rubber footwear empire in Woonsocket, Rhode Island, after the Civil War. He soon expanded across the border into Millville, Massachusetts. Irish immigrants honeycombed the area between the two states as this St. Patrick's Day flyer demonstrates, with participants from surrounding cities and villages.

ileges, and respect arrayed against the usual catch-traps erected by the Yankee ascendancy to curtail political power. The overwhelming foreign nature of the population also mitigated the influence of the old guard in the valleys of rural power and the corridors of the State House. Ironically, the Irish percentage waned in Woonsocket under the relentless immigration of French Canadians during the Gilded Age. However, they earned enough in salary to purchase sufficient property to vote under the antique provision of the state constitution that required until 1880 the ownership of $134 of taxable, real property. As early as 1874, Irish Americans formed an active chapter of the Adopted Citizens Free Suffrage Association, which lobbied to eliminate the financial qualification to vote and was later superseded by the Equal Rights Association. By 1876 the number of registered Democrats surged past 500 in the city, just a handful behind the ruling Republicans. The Irish-led Democrats eked out a close vote in the town that year in favor of liberalizing the state constitution to make it easier for immigrants to vote. They also endorsed a similar measure for Civil War veterans. Both issues lost statewide. Most French Canadians tarried in the area into the twentieth century before registering in significant numbers, even after rubber baron Augustus Bourn's constitutional amendment erased the real estate requirements for new immigrant voters in 1888, except in City Council elections.[16]

Occasionally incidents of intemperance, delinquency, and crime stung the Hibernian inhabitants of Woonsocket, but they seemed minimal compared to the situation in Providence.[17] Factory owners grudgingly closed their mills on St. Patrick's Day because Irish Americans skipped work to participate in ethnic festivities, especially the ubiquitous parades.[18] In addition, Banigan's Irish rubber workers made premium wages and enjoyed a lifestyle appreciably better than most of their compatriots elsewhere in the state. They would carve out a proud existence independent of other Irish-American communities.

Banigan also enjoyed John Holt's counsel from the beginning. After he arrived in Woonsocket in 1866 Banigan immediately "summoned"[19] his father-in-law to join him. Holt would spend the remainder of his life as the superintendent of the Woonsocket Rubber Company, practicing progressive labor relations. He also continued to use his rubber and chemical skills, inventing a rubber varnish for boots in the twilight of his career. Holt adopted the city as his own, becoming a philanthropist in his own right. He would play a conciliatory role between Banigan and his employees in the upcoming 1885 strike.[20]

Woonsocket Rubber Company: 1866–1884

In October 1869 Banigan purchased more land to construct another building for further expansion at a time when other local mills cut pay during the hard economic times caused by tariff problems in the woolen industry. Banigan felt a pinch during the mild winter of 1870-1871 and again in 1876-1877, when the demand for rubber footwear plummeted. The industry remained dependent on the weather for decades. Banigan dipped into profits again in May 1871 to purchase a specially built two-hundred-horsepower engine from the William Harris Company in Providence. By the end of the year the rubber company owned taxable property amounting to almost $90,000, the ninth highest assessment in the town. The first privately handwritten Dun (later Dun and Bradstreet) credit reports about Banigan in 1873 commented that "He owns some R[eal] E[state] but he has mortg[aged]. it for all its worth to one of the Banks." Still, the investigator found him "prompt and satisfactory."[21]

Banigan often left the day-to-day operations to John Holt, while he traveled around the country securing deals for imported rubber, new customers, and wholesale agents. In the spring of 1873, for example, he spent two months on the West Coast taking orders to provide specialty boots for excavators of precious metals. Several years later a former resident of Woonsocket, living in California, revisited her hometown and told the local newspaper, in an interview, that "through all the mining districts the Woonsocket pure gum rubber boots are usually worn by the miners; that she had at any time to look at the stamps on these boots to remind her of home."[22] Banigan left for European business trips on several occasions between 1874 and 1883. Holt also took similar journeys, including a voyage to the Continent. At different times they visited their respective homelands in Ireland and England. When Banigan returned to Rhode Island from a trip in 1873 the state was in turmoil. The Sprague empire, headquartered in the textile industry but with connections to a tangle of commercial activities throughout the state and nation, fell. The Panic of 1873 really originated here, but the ripples soaked the country and dampened economic activities in some sectors of Rhode Island for years.[23]

The depression that year also triggered a desperate attempt by local textile workers to obtain the ten-hour day, the crown jewel of labor's demands in that era when the workday usually stretched for at least eleven hours. Rhode Island employed almost 25,000 clothing operatives, about 5,000 of them under the age of fifteen. The state Ten Hour Association, although a paper tiger at the time, tried to orchestrate a general strike without the discipline required. Woonsocket, heavily industrialized, endorsed a call to action on 1 May 1873. The nexus of the walkout occurred in Olneyville, the production beehive located in Providence along the Woonasquatucket River. Despite the lack of any overt trouble in Woonsocket, Thomas Doyle, the Irish-Protestant mayor of Providence, sent a dozen police officers there to maintain the peace after textile workers abandoned the factories in large numbers. The strike petered out after several weeks without the requisite leadership of an organized labor union.[24]

Although the Irish now toiled in many cloth mills, the agitators behind upheavals in this period often came from Lancashire, England. The editors of the *Providence Journal,* always alert to a journalistic slap at the Irish, wrote that

> we cannot help asking ourselves how it is, that the Irish population in particular, who have told us so much of the oppression and tyranny of England practiced toward them in their native land, should, when they come to this free country, allow themselves to be led by a few of these same Englishmen who cannot leave behind them when they come to this better land, the terribly bitter spirit of the Trades Unions, whose practice has been more tyrannical than that of any existing government on earth.[25]

Factory operatives who sought a reduced workday finally got it with a vengeance: many plants put employees on half or two-thirds time because of the Panic of 1873, Woonsocket being no exception. The Bailey Wringing Company cut wages and the rubber company shut down for a time as well in February 1875. When it reopened in the middle of March, the very skilled bootmakers and "cutters" accepted a wage reduction, reflecting tough economic times. The predominantly female, less skilled, and lower-paid workers in the shoe department rejected any compromise, an indication that bargaining had occurred between labor and management. Most of the women, probably at the bottom of the pay scale already, refused to toil at an even lower rate, and walked out over a litany of similar issues like wages, working conditions, and gender that would explode a decade later in the 1885 controversy. During this first manifestation of labor turmoil, the firm employed approximately three hundred males and one hundred women who toiled ten hours a day, six days a week. Although we do not know the outcome of this strike, less than a year later, in 1876, the rubber concern reduced its monthly $15,000 payroll, along with almost every other manufacturer in the state. Those engaged in piece rate work lost one-fourth of their income while those on a set pay scale got trimmed 10 percent. A second visit from the Dun appraisers, written in abbreviations, provided an insightful look at Banigan but one that borders on ethnic stereotyping and may hint at Banigan's shrewdness during the personnel turmoil: "B. is considered good for any new undertaking," the agent wrote, but "has a Bombastic way that hurts him with the average of the people[—] a very loud talker [—]has been successful in his managt. of the Rubber Co wh. has been a profitable bus[.] has unbounded confid. in his own ability wh is shared to some extent by his bus associates, outside of this circle the confidence is not so unbounded."[26]

During this wrenching era of the 1870s the company announced plans to build a new, four-story structure, 50 by 160 feet "that would double the firm's capacity" now measuring an acre and one-half on a parcel of land at 44 South Main Street near the "falls." The rubber plant rented the first two floors to the Narragansett Horse Nail Company, another Banigan-Cook operation. As production grew to as many as 135 cases of boots a day (a dozen pairs to a case), a new generation of engines included ever-greater horsepower. The company added six new calenders to press the rubber, and each new structure housed a specialty in the footwear process. The enterprise operated wholesale warehouses in Boston and New York City. The peripatetic travels of Banigan and Holt paid dividends with global markets in places that would eventually include, before 1890, the American West Coast, Canada, England, Germany, Russia, Australia, China, and Japan. Banigan also introduced state-of-the-art technology to modernize his facilities for workplace comfort and efficiency. "Petroleum gas" would light the new building. The company's taxable property jumped to $143,000 that year and another $30,000 six months later.[27]

Banigan's vision outraced his ability to accomplish so many endeavors, but he kept long-range goals in place. In February 1877, one of the worst years of the depression and a time when the entire workforce was on furlough due to good weather hindering footwear sales, Banigan started one of his most important projects. He bought a warren of rundown mills and, more important, the water rights that went with them in the nearby town of Millville, Masachusetts, for $80,000. The immediate plan focused on redesigning the complex for a felting plant to produce material used in the lining of footwear. Banigan, throughout his career, eliminated middlemen and suppliers, preferring to run all facets of the production process under his own control and in close proximity to the main operation. In this case he bought out the Lawrence (Mass.) Felting Company and moved it to updated facilities in Millville. In 1881 he acquired the Hauten Sewing Machine Company for assembling boots and later the Hammond Buckle firm for adornments to his footwear.[28]

Banigan had another unannounced plan for the Millville area: he prepared to build the most modern boot factory in the world there while at the same time whipsawing Woonsocket against Millville for tax breaks. Banigan followed the Rockefeller oil scenario of vertical integration, soon going after the raw rubber itself, and eventually controlling the United States Rubber Company in the 1890s, while trying to horizontally integrate the competing firms. In the middle of that project, the Woonsocket facility displayed $275,000 worth of merchandise—competitively priced and of top quality—at the Boston Rubber Trade Show in June 1877, followed by another similar convention there in December. The company resumed full production in August. Oddly enough the Woonsocket Rubber Company still produced erasers for pencil tops on which Banigan held a patent and, perhaps out of nostalgia, a line of rubber stoppers and other sundries. These low end items must have had a small profit margin compared to a pair of expensive boots, but they highlighted Banigan's unwillingness at times to voluntarily give up any earnings that might pay bills during a downtime for some other line of rubber goods.[29]

Although Banigan continually upgraded the quality of the footwear through top-of-the-line machinery, sophisticated chemistry, and a

The Woonsocket Rubber Company, as well as most of its competitors, issued trade cards as a primitive form of advertising. (Author's collection)

highly skilled workforce, the process of making rubber boots and shoes did not change appreciably in the 1870s and 1880s. Other parts of the business would shift dramatically, however. For example, the American procurement of raw rubber from Brazil spiraled upward from 8 million pounds in 1872 to a whopping 25 million pounds in 1885. The price fluctuated as widely as the weight of the imported rubber: from fifty-one cents a pound to more than a dollar a pound in the same period, triggering a spirited contest to control the Brazilian export market. Two-thirds of the supply went to footwear production, the rest to a wide array of smaller specialties like those that Banigan produced on the side. But the almost assembly line structure of taking in rubber at one end of the operation and packaging footwear at the other remained intact, albeit only partially automated for another quarter of a century. The revolution in production was just around the corner, but only the farsighted could make out the developing outlines. The Woonsocket Rubber Company and the other handful of major producers scattered around southeastern New England in the Gilded Age employed 12,000 operatives by 1880, and their procedures were similar.[30]

Rubber Production

A reporter for the *Woonsocket Patriot* toured Banigan's plant with John Holt in 1877 and described the various steps that went into fabricating a pair of shoes or boots. "Cakes" of crude rubber, directly from the Amazon forest where legions of native gatherers collected the sap or gum of the wild trees, soaked in vats of hot water outside the factory. The square-foot pieces were black on the outside from a smoky hardening process used in the jungle, but temporarily remained white on the inside. Some of the blocks contained impurities from the harvesting procedure, partially inadvertent, and, as Banigan would claim on occasion, purposefully thickened with mud, debris, and other foreign properties in order to drive up the weight and cost. These accusations sometimes ended in legal action. The mixing room was a laboratory of chemicals used to transform the natural state of the gum rubber into a practical material. The usual suspects in this procedure included the distinctive-smelling sulfur, powdered whiting, oxidized lead, and coal tar. These elements, employed in very specific formulas, were mixed with the rubber at a later stage to create a more durable item with greater elasticity. The chemists always sought innovative ingredients to create better products, almost like a cook in a kitchen adding new ingredients to a familiar recipe.[31]

In another building, various machines shredded the rubber blocks with perforated rollers shaped like cylinders. Water spray during the operation dampened the friction and further purified the rubber. The rubber pellets were then matted and pressed into more manageable sheets and hung to dry in a separate room. Once dehydrated, the strips were subjected to a chemical stew and again passed through the rollers, several times, to set the thickness and expel any air bubbles. Lamp-black provided the final dark covering as the substance morphed from milky white to its opposite. Larger cylinders then adhered the rubber to felt or cloth lining which covered the different varieties of footwear: 6,000 yards a week at the Woonsocket facilities in 1877. More rollers embroidered and engraved the rubber sheets used for soles and heels. The human touch included a small army of craftsmen. "Cutters" cut the strips into the twenty-two sizes and configurations for one boot (or fourteen for a shoe) that produced the final shape. The cutters were "paid by the piece and work with lightening-like rapidity." The only children in the factory, usually teenagers serving an informal apprenticeship, cemented the soles and heels. The last operation required the sustained genius of Charles Goodyear: the footwear was varnished, hung on frames, and rolled into oversized ovens to be vulcanized by a dry, gradual, indirect heat. "When the rubber comes out from the oven . . . a marked change has taken place in its constitution," the reporter marveled, "it is no longer dead, but possesses a marked vitality." After further trimming, workers placed the boots and shoes in large crates.[32]

The economy in that era had its ups and downs, but at the end of 1878 the Woonsocket Rubber Company operated day and night to fill an overflow of orders as employment reached almost eight hundred. The one hundred operatives at Banigan's new felting company in Millville also toiled overtime. Observers from the Dun credit agency made several visits in this era, remarking that the company was "energetically managed," and "made a great deal of $ for the stockholders."[33] In reality, the stock was closely held by Banigan and his original backer, Lyman Cook, with only incidental paper in the hands of other investors. In order to protect life and property, mill workers trained at both properties to extinguish fires. Hydrants, hoses, and pumps guarded every entryway. Each floor of the mill could be flooded in seconds. Banigan hosted a day-long celebration for a new brigade at the felting company. The thirty operatives in the organization all had uniforms. Banigan later purchased a fire engine. Trained mill workers quickly extinguished a potentially serious blaze at the Woonsocket building in August 1881. The brigade also collected $100 for the widow of a fellow worker, and three hundred operatives attended the funeral. Such solidarity could be a two-edged sword as employees learned to do things in concert—both on and off the job—tucking away the experience for different purposes in different circumstances, like the forthcoming strikes in 1885.[34]

Banigan also voluntarily equipped his buildings with fire escapes, a safety feature not required under Rhode Island law in the 1890s. Both Banigan and Holt embraced another unusual corporate practice in that era, the indemnification of injured workers. One employee, hurt while operating a calender in January 1880, received full wages, and the rubber company paid for the physician. Holt, at the time, told of a similar accident when he worked at Bourn's Providence Rubber Company where he had received no compensation of any kind. In fact, Holt had his hand amputated in 1857 after it was caught between rollers on a calendar. His son lost three fingers in an earlier mishap. On other rare occasions when a worker died on the job, Banigan provided the widow with $1,000 and paid for the funeral. In a very unusual set of circumstances in 1883, the rubber company stopped production for four months, and Banigan remunerated half wages to the help as long as they returned when the company restarted, an astute move on his part as agents from other companies continually solicited his skilled employees to work at their rubber concerns. The almost-modern personnel practices seem sagacious today, but smacked of socialism in the eyes of shortsighted competitors and other Yankee manufacturers who had less of a bond with their operatives.[35]

In April 1880 Banigan and Cook, who was still the president of the rubber company owing to his initial investment, announced plans to build a new felting plant on a thirteen-acre site in Millville. In the back of his mind, Banigan knew this was only the start of a larger complex. He would construct the world's premier boot production plant and buy out Lyman Cook at the same time. In February 1882 Banigan purchased Cook's 1,250–1,300 shares (to add to his 400–600) in the Woonsocket Rubber Company for $250,000, the largest corporate transaction in Woonsocket history up to that time. Banigan paid one-third in cash and the rest in allegedly worthless stock in other companies. He also bought 100 shares owned by Cook's son. John Holt kept his 300–400, as did the treasurer, Francis M. Perkins who held 200. Rumors swirled about a force-out. When Cook died at age ninety in 1895 the *Woonsocket Patriot* wrote that "to the eternal disgrace of human nature be it said that men who were raised by the wealth and business influence of Lyman A. Cook to high financial positions played upon the nobility of their patron's honest nature and finally financially ruined their last and greatest benefactor." Whatever the details, the finger pointed obviously at Banigan in language that sounded eerily like a contaminated rubber case several years earlier. He now controlled the empire with unfettered authority to operate freely. He hired his son Patrick as a director. The Dun investigators, who called him bombastic a few years earlier, now stated that "Mr. Banigan has the full confidence of stockholders, is a shrewd capable man & has made considerable money."[36]

Stock soared to $250 a share. The city fathers of Woonsocket fretted over the new situation, fearing Banigan might move all his facilities across the border to Millville in order to consolidate operations and cut his tax rate. The General Assembly rechartered the company at a capitalization of $1.2 million. Banigan did relocate all boot production to Millville in February 1884. Upon the completion of new company housing, some bootmakers moved there. Others who commuted the few miles from Woonsocket began protesting about the cost of the train fare to the new facility. Inflation began to affect all aspects of life in shaky economic times.[37]

Banigan already had considerable experience in mill construction, going back to Jamaica Plain in 1866. He quickly decided to turn the new Millville building into a citadel for rubber boot production while simply making improvements to the felting company. He followed the recommendations of the insurance industry to the inch. Banigan fabricated eighty tenements to house the new army of workers who would overwhelm the scarce living quarters in Millville, and also built a schoolhouse. He equipped the new plant with telephone service to the

Woonsocket factory. The facility sported the latest in elevators, including an unusual safety switch for employees. Two-thirds of the rubber company's biweekly payroll of $45,000 went to Millville by late 1884. The enterprise featured the largest mixer in the industry at a width of 54 inches. The United States Rubber Company underwrote a secret appraisal of the plant in 1894 by future president of the cartel and fellow Rhode Islander, Samuel Colt of Bristol. Almost a dozen years after construction of the facility, Colt could write, preparatory to purchasing the Banigan empire, that the Millville works were "unquestionably the best constructed and most complete for the economic manufacture of rubber boots of any factory in the world."[38]

Banigan's boyhood history provided little evidence of future success: his humble birth on the Shirley Estate in Ireland; the tumultuous and controversial life of the Irish in Rhode Island when his family arrived; and the rubber king's adolescent neighborhood in Providence's Fifth Ward. In some ways, Banigan soared above it all but could never quite escape the gravitational pull of his ethnicity and religion. His return to a more insular Rhode Island in 1866 to run the Woonsocket Rubber Company, after an exciting smorgasbord of opportunities in the more tolerant outside business world of science and production, forced him to face the shibboleths of narrow Yankee dominance once again. Massachusetts, while rent with similar discriminatory trends against the Irish and others, did not institutionalize second-class citizenship in its constitution. Banigan, this time around, did more than face the prejudice, he faced off the discrimination. He reconnected with his roots and seemed to symbiotically join them to his considerable commercial success and experience. During the Woonsocket years he built the most sophisticated footwear establishment in the world at Millville and bought out his investors to unilaterally take the helm of his expanding domain. He slayed creditors, doubters, bigots, and competitors alike. Then in 1885 he faced an enemy from such an unexpected quarter that he almost self-destructed in his wrath.

The Strike Setting: The Knights and the Federation

Labor activity in the 1870s was spasmodic and ephemeral both locally and nationally. Banigan witnessed union activity first-hand in Woonsocket in the 1870s when agitation for the ten-hour workday spread among the general workforce. He also experienced a brief strike by his own employees against cutbacks in 1875 caused by an enduring depression, although not much is known about that walkout. However, the roots of the Rhode Island and Massachusetts labor movements were as old as the industrial revolution. The Mechanics' Associations in Rhode Island, with agitators such as Seth Luther at the helm, led the charge for expanded voter rights even before the Dorr War in 1842. Although the first major flowering of unions featured both practical and idealistic goals, the Gilded Age manifestations sometimes embraced a narrower craft outlook.

A handful of strikes that galvanized public attention to the contemporary "labor problem" offered a glimpse of the chasm between workers and management in the last generation of the Gilded Age. Railroad workers initiated a violent walkout in 1877 that rattled the country, while miners punctuated the stillness of economic stagnation on several occasions. The state of Pennsylvania hanged twenty Irish coalminers, alleged Molly Maguires, who practiced sabotage and murder on the job. Workers, at the time, lacked an all-encompassing organization that could do battle with captains of industry who increasingly hardened into robber barons. But help was on the horizon.[39]

Beginning in 1869 a clandestine union movement grew from a handful of tailors in Philadelphia into the nation's premier labor organization when it went public a generation after having trained thousands in the culture of concerted working-class activity. The Knights of Labor, at first operating almost as an underground group, eventually practiced an unprecedented inclusiveness of the skilled and unskilled: black and white, male and female. They also offered to arbitrate matters of labor relations, although not too many employers accepted that offer until the Order defeated the most notorious capitalist of the day, railroad owner and financier Jay Gould. The Knights' labor battles spilled out from worksites to neighborhoods; from bargaining sessions to ethnic and blue-collar cultural events; from second-class deference to political mobilization. The movement spread like wildfire in justice-thirsty Rhode Island. The Knights, under the leadership of Irish-American Terence V. Powderly, brought industrial civilization to factories mired in medieval practices: grievance procedures, day care centers, education and child labor laws, and a temporary but imperfect ten-hour law. The Noble and Holy Order of the Knights of Labor took on the trappings of a religious revival in the state and, while the cast was diverse, Irish-Catholics played a dominant role.[40]

In neighboring Massachusetts, where political democracy trumped Rhode Island's wayward practices, the Knights flourished under the leadership of some of the greatest personalities in the movement at the time. Racial, ethnic, and gender equality spurred the Bay State Knights into becoming the largest state affiliate in the nation. Workers responded in a militant fashion, striking so often and freely that unstructured actions eventually undermined the ability of the Knights to hold the empire together.[41]

While Banigan's operatives would choose the Order to represent them, another labor organization barely visible at the time would loom large in the country's future and represent some of his employees later. The Federation of Trades and Labor Unions, formed in 1881 almost as an annual debating club, would morph into the more familiar American Federation of Labor in 1886 under the leadership of Samuel Gompers. Like the Knights, the Federation served as a seedbed for future, influential labor leaders but with an emphasis on craft unions. And part of its early focus eyed liberal Massachusetts as a dispenser of progressive labor relations to the surrounding area. The group allocated $100 from its meager resources to investigate conditions in backward Rhode Island. Two brilliant organizers from Massachusetts, Frank K. Foster and Robert Howard, born in Lancashire, England, reported widespread child labor throughout the state, work weeks that stretched to seventy-four hours, and the failure of ten-hour legislation. The condition of textile workers cried out for a solution so badly that the infant, threadbare Federation organized rallies in Rhode Island three times in 1883, including two ten-hour demonstrations in Woonsocket. The American Federation of Labor was only a rising star at this time while the Knights' meteoric rise illuminated the industrial landscape immediately. Some labor unions gathered around a singular trade instead of embracing the amorphous notion of working-class solidarity that earmarked the earlier workingmen's parties in the 1830s as well as the Knights. Bricklayers, carpenters, and machinists, to name a few, flouted their craft identity as skilled workers not easily replaced by others. Although some Irish made inroads during this period, their entry into the ranks of labor usually earned them only a temporary, second-class citizenship: helper, tender, or carrier—laborers in training assisting talented tradesmen. However, over the next generation Irish workers came into their own, especially as craft unions eventually supplanted the Knights. Coming from a nation where antiquated British mercantilism curtailed most economic development and labor activity, the Irish in the United States and New England embraced the concept of unity among skilled colleagues and employed it for financial advancement initially in the Knights' Order.[42]

Banigan and his workers, despite any underlying antagonisms over personnel issues, operated within a hostile Yankee milieu whereby they gathered strength by sticking together as a persecuted minority in a twisted kind of camaraderie that transcended economic class. The *Woonsocket Evening Reporter* observed "a monopoly of unusual favoritism in the nationality employed" at the rubber company, although the workforce included other nationalities because Banigan publicly endorsed diversity. He retained the loyalty of his Irish-American workforce through hiring policies and a high profile activism on both sides of the hyphen. On several occasions he threatened to tackle the Republican machine and run for state or national office. The *Providence Visitor* explained his fervor: "It was generally believed there was additional zest to the competition he encountered because he was an Irishman and a Catholic." The *Boston Citizen,* on the other hand, an anti-immigrant newspaper, maligned Banigan as "an Irish Roman Catholic of the most bigoted type" who allegedly later replaced Protestant rubber workers with his own kind at another Rhode Island factory. "No one can get employment there now," an article complained, "unless he drinks holy water and crosses himself every time he turns around." By 1885 French Canadians pushed the Irish to second place in demographics in Woonsocket with 4,811 inhabitants in a population of just over 16,000. Despite political limitations in most parts of Rhode Island, here the Irish community enjoyed stable employment and good wages, thanks in large part to Banigan's policies. The Irish in Woonsocket were politically active and accounted for just over 500 eligible voters compared to almost 1,100 American-born citizens, a number that probably included many of Hibernian ancestry. Still the numbers reflect the state's restrictive voting laws, which limited the franchise even when industrial workers earned a decent living.[43]

Although Irish nationalists often made pilgrimages to Banigan's home (and even to his father-in-law's house in Woonsocket) on trips through Rhode Island, Banigan's ethnic patriotism took a back seat to his religious partisanship. Trying to seperate Irish nationality from the Catholic religion is a difficult, delicate operation even within that group or among its enemies. Having arrived in the United States at the tender age of eight, Banigan's memory of the Catholic Church in County Monaghan must have been hazy at best. However, the American church, replete with mostly Irish priests, took up the slack and remained a continuing, concrete force in daily life. Banigan's brushes with death in a train and a ship wreck may also have strengthened those religious tendencies. Similarly, he correctly viewed the local church as socially active. His investment in various charities provided almost immediate, visible results whereas endowments in Irish causes abroad always carried a long timeline that might be punctuated with violence. The lion's share of his legendary generosity went to various Catholic agencies, although he participated in Irish groups at both ends of the social scale.

Just before the 1885 strikes, Pope Leo XIII announced that Banigan had been knighted in the Order of St. Gregory, an honorary post in the Pontiff's personal honor guard that recognized outstanding Catholic philanthropy. Banigan was only the second American so honored. Ironically the letter of announcement was written by Cardinal Simeoni, head of the Vatican's Congregation of Propaganda, who also had proscribed the Knights of Labor a year earlier. In many ways, Banigan's award was an accolade for the state's entire Catholic population, although the Yankee press basically ignored the recognition. His employees seemed as dedicated at the grassroots level of religious devotion as Banigan was at the acme. Although there is no way to determine the symbiosis between titan and toiler on these issues, workers contributed generously through weekly collections for parishes and the diocese. Woonsocket papers reported innumerable special appeals at the local St. Charles Church (established in 1844) that yielded impressive sums. Several surveys of Catholic Church attendance in Rhode Island at the end of the Gilded Age showed rates close to 90 percent.[44]

Rhode Island, the Blackstone Valley, and the Woonsocket-Millville nexus had long been hotbeds of Irish nationalism and Irish-American political and cultural activity. Groups included the Ancient Order of Hibernians; Sarsfield Literary Association, Irish National League (which met in the basement of St. Charles Church); Clan na Gael; one of O'Donovan Rossa's revolutionary skirmishing clubs; and Father Mathew Temperance Association whose president Joseph McGee was secretary of the local Knights of Labor. Father Theobald Mathew had made a legendary visit to Woonsocket in 1849 and that anniversary was always noted with great fanfare. Area newspapers, meanwhile, provided space to Fenian affairs domestically and abroad. The Rhode Island Land League (a movement to transfer farming property in Ireland from landlords to peasants) at its 1881 outing in Warwick— about twenty-five miles from Woonsocket—featured 250 participants from the region, the largest contingent at the event. Between 1880 and 1881 the Irish in Rhode Island donated more than $7,000 to the group, the fourth highest total in the United States behind Pennsylvania, New York, and Massachusetts. A few years later the local section of the League refused to funnel its substantial offerings through the state branch because "the patriotism of Providence Irish-Americans is only spasmodic, that of Woonsocket Irish-Americans is steady from day to day and year to year." Reports of visits to the "Auld Sod" by local inhabitants, high and low, kept communications with the homeland fresh and personal. Irish names also filled ballots during Democratic Party caucuses as they slowly pushed the GOP from power in former Yankee strongholds, especially in the state's urban centers after property qualifications were dropped in 1888 except for city council races in Rhode Island's five cities. The wide array of groups, embracing many sophisticated political agendas, intensified the ethnicity and knowledge of the Irish while providing valuable, concrete organizing abilities. When the showdown with Banigan came, his employees seemed confident and secure from their participation on so many training grounds with radical agendas. Taking on the boss did not seem so monumental for those skilled in debating, raising funds, arguing about economic problems here and abroad, or mustering volunteer soldiers.[45]

A local Baptist minister who witnessed the influx of Irish and French Canadians into Woonsocket said without rancor: "This is a Catholic town." The Irish had built the foundations of the Church, literally and figuratively, by dint of their early arrival in the 1830s. The original pastors, like most of their counterparts throughout the United States, placed a heavy emphasis on the social and political Americanization of the Irish flock as the means to economic success and social acceptance. Most area priests at that time were Irish-born and embraced acculturation but often encouraged parishioners to retain an Irish heritage. One Woonsocket pastor even led an attempt to revive study of the Gaelic language. Despite the Church's official opposition at one time or another—and at different levels of the hierarchy—against the Knights, Land League, Fenians, and Ancient Order of Hibernians, the groups flourished anyway.[46]

By the time of the 1885 strike Banigan had nurtured his firm into being the premier producer in the field with a sterling reputation for quality footwear. He controlled one-fourth of the market and refused to sell "inferior grades" like his competitors around southern New England. The Woonsocket Rubber Company was the second largest manufacturer of boots and shoes among the region's seventeen firms. The other concerns would understand the Banigan way in even greater detail when the United States Rubber Company formed in 1892. He continually employed new technology, recycled, bought out his suppliers, and practiced progressive health and safety procedures. He understood the connection between modern, safe work places and increasing profits. By the 1880s he had created a global market for his footwear. Similarly, his labor-management practices remained progressive if paternalistic until the walkout. Banigan's work environment would have been welcome relief at many establishments, while his Irish hiring policies conspicuously set him apart from almost every other large Rhode Island employer who often stuck a "No Irish Need Apply" sign at their front door.[47]

Banigan's public relations were as much a proven process as his rubber production. He practiced a form of social justice laced with paternalism at the factory and in the community. He often donated large sums of money to local relief agencies in tough economic times. During the vicious Panic of 1873, which hit Rhode Island particularly hard, Banigan refused to cut wages as other manufacturers did. He also eschewed most child labor, in a state that was the most notorious abuser in New England at the time. Of the 300 rubber workers identified in the 1880 Woonsocket street directory, 118 owned their own houses. There was also company lodging in both towns where an estimated two-thirds of the employees resided. One local newspaper claimed that Banigan's workers had "snug sums in the bank" and financial investments to go along with home proprietorship. Individual savings accounts in Woonsocket—on a percentage basis—were the second highest in the state with an average deposit of $492.68. The Irish had already come a long way from rent rolls and peat fees back home.[48]

Banigan continued to steer his rubber works through a series of shaky financial situations. Although Rhode Island eventually rebounded from the Panic of 1873, the recovery was tentative because another recession hit a decade later. Woonsocket police noted more "tramps" than ever before. The *Providence Journal* lamented the recurring commercial nightmare in 1885: "The hope so generally entertained . . . that the new year would dissipate the depressing dullness which had already too long prevailed, has not been realized. On the contrary, matters have, as a rule, gone from bad to worse." The *Providence Visitor,* the weekly newspaper of the Catholic Church, set a protective agenda for its wage-earning parishioners. "There seems to be an unholy conspiracy on the part of the manufacturers of this State to scare their employees by reducing wages," an editorial warned, "but we must earnestly call on the workingmen to form unions and to support them so as to be able to meet those who thus proclaim their hostility." The editors did not anticipate that the largest labor-management struggle that year would involve its own members at both ends of the work floor. Still, Bishop Thomas Hendricken, just before his death in 1886, wrote a complimentary letter to the Rhode Island Knights of Labor: "I who have lived in factory towns for thirty-three years say Godspeed to the Order." This valedictory was a far cry from the antagonism and damnation emanating from many quarters of the Church's hierarchy.[49]

A KNIGHT OF ST. GREGORY
AGAINST THE KNIGHTS OF LABOR

"Mr. Banigan, who turns out the best goods in the country,
is himself the smartest rubber man in the world but with
all his generosity and enterprise, has lost an opportunity
to allow us to place the crown upon his head."
—*Striking bootmakers at the Millville plant, organized*
into the Knights of Labor, July 1885

"I made promises to the Millville bootmakers which
Mr. Banigan will not sustain. He tells me I have no right
to make promises, as I am not running the works. . . . If I
were running the works I could have settled the matter long
since. . . . I only hope Mr. Banigan has done wisely for
himself and the community in refusing to endorse the
settlement which we thought was made, and which
would have been satisfactory to all."
—*Fr. Michael McCabe, Catholic vicar general,*
acting as a mediator, October 1885

The rubber footwear business fared badly in a recession that lasted into 1885. Banigan complained that "our competitors can undersell us and drive us out of the market." Many businesses in the Blackstone Valley that technically stretched from Providence to Worcester were already paring back wages as sharply as cutters pared soles and heels. On 2 February, Banigan unilaterally announced an average 18 percent wage cut about the same time one of his former partners went bankrupt in a different venture: "We found that we had been paying more than any other company in the country." Similar operations in the region quickly followed suit. Skilled operatives, miffed at Banigan's one-sided decision to reduce wages without any forewarning, struck as soon as notices about the cutbacks were posted.[1]

Some 325 skilled bootmakers and rubber cutters, the elite of the workforce led the spontaneous walkout in Millville. They earned between $2.00 and $3.00 a day owing to piecework incentives during a 60 hour week. They literally stood in the middle of the production process with machines doing an increasing portion of the labor before the product reached their work stations. After vulcanization less skilled handlers trimmed the boots and shoes. Banigan liked to boast that he scientifically streamlined the process so bootmakers and cutters did nothing but their own specific jobs and made more because of it. For example, semiskilled operatives performed the tedious chore of cementing rubber strips to footwear, "thereby enabling them to use the time devoted to this work in other mills to increase the capacity of their daily work in ours." Banigan claimed that other unskilled workers carried parts directly to the benches of the craftsmen in a kind of moving, human assembly line "thus avoiding any necessity for losing any time in procuring them."[2]

Skilled workers challenged some of Banigan's rationalization moves. Fierce competition in the industry, they complained, triggered spiraling fabrication quotas that now forced eight cutters to squeeze into work benches designed for six. There was no longer enough room for comfort or any place for on-the-job philanthropy at the point of production, that mystical spot or operation in a mill that first transforms raw material into a profitable commodity. The evolving process of manufacturing and cutthroat wholesale and retail rivalry poisoned even sincere paternalism. Banigan sliced the lucrative piecework rates by 15 to 25 percent for cutters and 12½ percent for bootmakers. Several years later an industry journal explained the rubber king's Darwinian system: "By a process of selection, keeping the most expert and eliminating the less expert, and by employing opportunity to develop to the greatest extent possible the special aptitude of the workman, a body of the most skilled boot makers living were attracted to Millville, to a factory which had the reputation of running the year round." Banigan automated his existing facilities as fast as technology developed and built several highly mechanized new plants a few years later. However, boot production, the most complex and labor intensive process in the footwear business, was not so easily transformed, allowing skilled cutters and bootmakers to avoid for a time the eventual slippage to the ranks of the semi- and unskilled.[3]

Strikers met almost immediately with Superintendent John Holt to inquire "if their former wages would be restored as the times grew better." Although wages dominated the initial conflict, the issues proliferated as the walkout grew bitter. Holt replied that the firm needed to equalize pay scales with rivals on a permanent basis, and then asked them to return for a few hours to finish the work left behind when they walked out. Amazingly, they reported to their benches to complete their assignments so as not to saddle Banigan with the cost of spoiled rubber that rotted when left to the elements. Employees then went back on strike the same day, a fascinating example of the period's waning producer ethos, where workers acted as guardians of the workplace and its material and equipment. They held a number of meetings to discuss the situation, especially in Woonsocket where many lived and could be accommodated at a larger venue. They elected a strike chairman who was also head of the local athletic association. The spokesman emphasized the wider importance of the walkout: "The matter is one which affects not only ourselves, but the whole business of the community. Be not hasty in your actions." The skilled workers agreed to another session with management (Banigan apparently was away on business at the time). A committee tried to negotiate a full-time work schedule to replace seasonal employment or else set a date to reinstate the old wage rate. One veteran hand complained bitterly when he refused to accept any reduction in his substantial salary: "I never cut for less than $3 a day, and I ain't going to begin now."[4]

Two weeks into the strike, John F. Holt, the popular superintendent, rubber master, Banigan's father-in-law, and an English Protestant convert to Catholicism met cordially with the employees again. Holt, who had been with Banigan since the inception of the Woonsocket Rubber Company, gave fraternal advice to his respectful audience. He underscored the anemic economy but maintained that skilled workers at the Woonsocket Rubber Company remained the most affluent in the industry. He reminded recent Irish immigrants that the average $60 a month was a princely sum. Holt also chastised them for ingratitude after the company distributed $30,000 to unemployed rubber operatives just two years earlier at the beginning of the recession. He

An engraving of John Holt, Banigan's father-in-law and mentor. Holt was a rubber master in his own right at the Woonsocket firm who earned the respect of his employees. He was an English immigrant who converted to Catholicism and donated heavily to charitable causes. (The portrait is in Richard Bayles, History of Providence County *(1891) 2 vols., 2:402)*

decried their actions in locking out 350 lesser skilled, day laborers in the calender room, varnishing department, and packaging center although they too, unbeknownst to him, would be organized within a year. Most workers sensed the need to stick together. Holt reflected the era's obsession with employment as a sign of masculine virility: "walk forth like men if you do not wish to return to work and thereby avoid the curses of the mothers and the children that you are depriving of bread and keeping in a starving condition." But despite a generation gap he also promised there would be no recriminations. Old-fashioned paternalism strained to accommodate the fledgling mass production milieu and the industry's brutally tight market competition. That scenario transformed the Knights of Labor into industrial unionists in the Woonsocket-Millville rubber arena as they reluctantly embraced an emerging way of life devoid of what was left of *noblesse oblige* on the job.[5]

After Holt left the meeting, workers voted to allow reporters to stay if they refrained from identifying individual strikers. Employees cited the firm's wealth and condemned the mechanized drift into total piecework whereby dies, mallets, and machines replaced the keen eye of bootmakers and cutters. On the other hand, the usual argument that a half a loaf was better than none resonated among the crowd in the adverse economic climate, as well as lamentations that the group had acted spontaneously before organizing a union. Part of the confusion was an honest inability by either side to determine exactly the comparative wage scales between different firms in the rubber industry. Banigan adamantly claimed that he paid from $\frac{1}{2}$ to $4\frac{1}{2}$ cents more per pair of boots in labor costs, depending on style. However, even he eventually lowered his estimate of the differential.[6]

Bootmakers, conversely, carped that they were now the lowest paid in New England, an exaggeration on their part given the state of several footwear producers who ran slipshod operations and sold inferior goods based on low wages. They also pointed out that an extra strap had to be fastened on Woonsocket products, adding yet another step to the dozens of parts and pieces that constituted a solitary rubber boot. The employees also complained that it was hard for them to obtain data from other reticent rubber firms "because manufacturers have a strong union." Eventually the battle turned esoteric over whether the extra piece was a metallic loop, a ring, or a web strap. A published comparison of five different companies cited eighteen different styles. After one bargaining session the Knights confessed bewilderment at Banigan: "his propositions were so cunning and complicated and the contrast between the system by which other concerns are run and that, by which his is run, that to bring about a settlement on conditions offered would be an impossibility." Banigan would employ similar legerdemain almost a decade later in negotiations with the United States Rubber Company. Later in the walkout a local priest, Father Michael McCabe, personally visited a rubber plant at Malden, Massachusetts and viewed a disassembled boot that had the same number of pieces as the Woonsocket Company, not fewer, contrary to the strikers' estimation. With so much confusion and a lack of organization, they voted halfheartedly to return to work after three weeks "in the interest of small stockholders, and themselves, as well as the public." They vowed "to await a more favorable time for action" and raised $200 for the next engagement with the company. Ironically, and a reflection of poor planning by the strikers, the firm was about to undergo its annual month-long layoff that was often repeated several times a year depending on weather conditions and the resulting exigencies of supply and demand.[7]

The Knights of Labor

Joseph Banigan met with employees at the Millville facility on 7 April 1885, six weeks after the original walkout and while they were still furloughed. He announced that his rubber company would pay as much as any other concern if not higher. Banigan then asked how many had joined the "Union league." The show of hands was almost unanimous, a courageous and collective expression of emerging solidarity among the rank and file. Banigan allegedly said: "I will neither act with nor receive any committee from the union. I will shut down my works. . . . I will never, NEVER, yield." He also taunted his employees by insisting that "traitors" among them had already tipped him off about the organization. Banigan reacted to the labor action with surprise, pain, and arrogance that mimicked the landlords and superintendents of the Protestant ascendancy in Ireland and the industrial barons of the Yankee ascendancy in Rhode Island. He was about to cross the line. The first local assembly (LA) of "rubber factory hands" in the Knights of Labor had materialized at his Woonsocket plant three years earlier in 1882 as LA 2297, two years before the next exclusive assembly of rubber operatives formed in New Jersey. The Knights organized the Woonsocket plant without any apparent opposition or public controversy, probably under the leadership of the accomplished bootmakers. That same year Banigan relocated that end of the business to Millville, a transfer that was some time in the making and included whipsawing the two towns for tax concessions. Hard feelings in Woonsocket over the controversial move during a recessionary era when many other local factories were closed helped the strikers garner public support across class lines when the action resumed in June. The bootmakers were apparently a part of the original formation of the rubber workers in 1882: "We were organized in Woonsocket as a local assembly of the K. of L. attached to the general assembly [in Rhode Island] but as our factory was [later] located in Millville we came under the jurisdiction of District Assembly 30 [in Massachusetts]." The concentration of skilled bootmakers at Millville within another state left the less skilled, heavily female, mechanized shoemakers to fend for themselves for the time being in Woonsocket, even though they belonged to the pacesetting, initial LA in the rubber industry.[8]

On June 29, 1885, almost three months after Banigan asked his employees how many belonged to a labor organization, rubber operatives at Millville, now "secretly" affiliated with the Knights of Labor, quit work at ten o'clock in the morning after a prearranged whistle signal. Unlike the first walkout in February, skilled cutters refused to return and save $10,000 worth of unfinished rubber that might spoil when exposed to air. Banigan called the action malicious and vindictive. He threatened to sue but admitted later that he was able to preserve most of the threatened stock. The Knights claimed that they had hurt their own cause when they completed material the first time around and, just as Banigan had given no notice when he unilaterally instituted the original wage reduction, so they were now repaying him. The operatives' producer ethos was evolving into hardball unionism. They abandoned an old system of workplace relations and responsibilities but only after management took the first step in the changeover: "What Mr. Banigan calls malice in us we term shrewdness in him." The quest for the old wage level drove the new strike, but Banigan's dismissal of two employees—both officials in the newly formed Millville

Knights of Labor LA 3967—triggered the actual job action. Banigan disingenuously denied knowing they were officers in the union. Workers had met regularly since the initial walkout and prepared to square off in a strike that would stretch into October and raise issues far removed from the narrow battlefield of labor-management relations. The United States Commissioner of Labor called it the nation's first strike by a *union* of rubber workers.[9]

The Knights of Labor had been a clandestine group from its birth in 1869 until going public in 1882 under the dynamic leadership of Terence V. Powderly, a machinist and mayor of Scranton, Pennsylvania. Historians of the organization have been stymied in interpreting the Order, because it seemed to be all things to all people, at least for several years when it engaged in industrial warfare against Gilded Age employers. The organization featured many contradictions between the leadership and the rank and file. One thing for sure, they sensed in the words of sociologist Kim Voss, who wrote about the order in New Jersey, that a critical "moment of working class formation" had arrived. The economic downturn, volatile personnel issues, increasing automation, and the emergence of a powerful union set the stage for one of those rare moments in industrial history. The Knights' craftsmen in New Jersey and, in this case, the rubber artisans in Massachusetts, encouraged their less skilled co-workers to form assemblies as well. Elsewhere internal tensions between workers with different levels of proficiency caused jealousy and interfered with union solidarity.[10]

In Massachusetts the Knights of Labor, DA 30 became the Order's largest national division. In Rhode Island, the first urbanized and industrialized state, disenfranchised workers sought basic rights and conditions within a stifling framework of political control by factory owners. With a less democratic foundation than its neighbor, the Rhode Island Knights nonetheless grew rapidly in the 1880s, helping to expand voting privileges, to establish a labor commission, and to enforce school attendance laws. They also founded a day care center and worker cooperatives.[11]

When operatives struck Banigan for the second time in 1885, the conflict offered an unusual look into the minds of a manufacturing titan and his toilers. A full-time breadwinner since the age of nine, Banigan pulled no punches: "I have worked as hard as any man employed at the works, and I will roll up my sleeves again sooner than allow employees to take the reins in their hands." He was so used to doing things himself that he acted as his own spokesman. His pronouncements to local newspapers during the walkout were personal and refreshingly candid, not prepackaged statements by a corporate lawyer. The dialogue during the walkout provides some of the best primary source material about Banigan's life and outlook. He argued to influence public opinion, especially the Irish-American community. He could not escape his humble beginnings. Furthermore, he expected his own work habits and success to rub off onto his now ungrateful employees as a further manifestation of his own magnetic influence.[12]

Ironically, his countrymen at the factory had indeed adopted his pride in craftsmanship and brash demeanor. They turned the tables on him. Workers spoke openly with the press and the public as well, providing another rich body of working-class testimony usually unavailable. Within a couple of days of the second walkout, the Knights of Labor held a rally in downtown Woonsocket. They marched with a band to the railroad station for the short trip to Millville and a larger demonstration by 1,500 labor partisans. Joseph Hines, the discharged master workman of LA 3967 at Millville, presided. He was a two-year veteran cutter at the rubber works, lived in company housing with his widowed mother, and was also the secretary of St. Augustine's Temperance Association in Millville. Allegedly he had been fired for production mistakes and for overlooking an imperfection in someone else's work, a new source of irritation that marked the gradual introduction of techniques that turned some workers into inside contractors and competitors. Hines placed the responsibility for the few errors on "bad stock," a complaint Banigan often registered about crude rubber himself. The Knights agreed that some thirty union members had been dismissed justifiably from the Woonsocket Rubber Company for shoddy workmanship since the formation of the LA in the spring. The Order had not protested these firings but felt Hines' termination was unwarranted and due solely to his position in the Knights.[13]

Besides the industrial quarrel over wages and a union, workers also complained about disrespectful and vulgar superintendents who also seemed to feel the pressure to produce more footwear at an accelerated pace that strained the former goodwill between shop floor management and the rank and file. The union demanded that "a more gentlemanly conduct" replace cursing and harassment on the job. Management usually leveled charges of indecorum and crude language against workers, not the other way around. Here the operatives took the high road, capturing public and religious support, especially that of the Catholic Church. Formerly, employees regularly presented foremen with gold watches and other gifts on special occasions and anniversaries. Now supplications to local citizens and merchants fueled a boycott against a handful of scabs and supervisors.[14]

Workers at Millville cemented an alliance with other rubber workers to institute a primitive master contract through overlapping strikes. Historically, these footwear operatives, similar to textile employees throughout New England, moved from town to town in search of better conditions, improved wages, or even a spouse. During seasonal shutdowns families visited former colleagues. Some moved to escape debt. Others found promotion into managerial positions after gaining experience elsewhere. In a touching example of solidarity, rubber operatives at the Para enterprise in South Framingham, Massachusetts, sent a memorial pillow to the funeral of an Irish woman and former co-worker at the Woonsocket plant, inscribed "shopmate." They formed a primitive, vocational network that also operated as a personalized, regional grapevine.[15]

Banigan, when asked at the outset about the hostilities, flippantly replied: "You see there is an organization known as the Rubber Workers' Union, or Knights of Labor, or some such society." However, he studied the union's constitution (probably provided by one of the "traitors" he mockingly mentioned earlier) and later changed his characterization by saying the Order was too smart for such an ill-conceived action. He also portrayed his skilled cutters and bootmakers as hot-headed boys "16 or 17 years of age," lucky to earn such a good living, although his characterization of their youth was probably exaggerated and any negative references to their abilities unwarranted.[16]

Despite the visible anger at the demonstration, the rally remained peaceful although Millville had no police force. The rhetoric heated up after that. Joseph Hines, the dismissed president, introduced key leaders from the Massachusetts Knights of Labor who were in turn influential on the national scene. Albert A. Carlton, District Master Workman for the entire Bay State and the titular head of the leather shoemakers' union, pledged total assistance. Charles W. Hoadley, a rubber worker and president of the Knights' local at the Para Rubber Company in South Framingham, also spoke as a representative of an emerging regional network of rubber footwear producers within the Order. The Knights chartered eighteen rubber assemblies, fifteen of them in 1885 and 1886 during the Great Upheaval, covering most of the significant footwear factories in the tri-state area of Massachusetts, Rhode Island, and Connecticut.[17]

The group apparently had coordinated other recent rubber strikes including one at the Para plant and another in Bristol, Rhode Island, for 1 May. The walkout at South Framingham had limited the wage reduction there to a set period while also setting up arbitration procedures to avoid future conflicts, an important development as strikes skyrocketed out of control in the Order. Millville held out longer, trying to restore the pay scales completely, as a triumph over Banigan and the pacesetting Woonsocket Rubber Company would weaken the resolve of other owners. Hoadley then expressed the rippling power felt by the Knights in 1885: "we broke the back of that arch-millionaire, Jay Gould, and we will show this pygmy [Banigan] how quickly we can wind up his little ball of yarn." The Knights must have been feeling their oats as they permitted his secretary to sit on the platform and take notes but would not allow the rubber king to address them. The reference to the Gould struggle was only the first of many to cite that signal victory.[18]

"From Cellar to Attic"

The strike at the Millville plant was suffocating. Almost the entire workforce of one thousand stayed out. Some fifty employees who remained at work the first day quickly dwindled to a dozen, flushed out by striker and community pressure. The cutters and bootmakers realized the need for unity between skilled and unskilled and organized the whole mill "from cellar to attic" soon after the strike. Still the Knights seemed to regret the necessity of a walkout: "Mr. Banigan, who turns out the best goods in the country, is himself the smartest rubber man in the world but with all his generosity and enterprise, has lost an opportunity to allow us to place the crown upon his head." The Knights lamented the recriminations against them after the February imbroglio, "when tyranny was carried to the most extreme point like the landlordism and eviction in Ireland." They unflatteringly portrayed Banigan as an Anglo-Irish turncoat who took his peasant-workers for granted, the kind of barb that must have deeply pained him. Throughout the conflict loose references to the Irish character by both sides and by local newspapers as well discredited both the rubber king and his employees. Each understood the acidic historical analogies and the vulnerabilities of the other.[19]

With the Millville plant shut tight, the combatants cranked up sophisticated public relations campaigns. Banigan let fly with a letter of appeal that appeared on the front page of the *Woonsocket Evening Reporter* and showed up in other newspapers as well. He apologized "for encroaching upon your valuable space, but I desire that the public shall know the facts, just as they are." He honestly decried the turmoil because "I have always felt great sympathy for the workingman, and have always endeavored to do all within my power to elevate and improve his condition." He then explained the need for wage parity with his competitors, argued that pay reductions were insignificant, and described the superior working conditions at his plants. He dismissed allegations that company supervisors were abusive in any way and enumerated why thirty-two workers had been dismissed in the last eighteen months: twenty-four for poor work; four for drinking; one for fighting; two for disobedience; and one for meddling. Banigan's survival-of-the-fittest process continually and purposefully terminated the less skilled just as Evelyn Shirley and his agents had weeded out the weaker tenants on the Irish estate in County Monaghan during the Famine.[20]

Banigan reiterated the reasons for the dismissal of Joseph Hines, president of the Knights' Millville assembly, for inferior work, not labor activity: "We care not to what trades union or organization our men belong, nor do we care what their creed or nationality may be," a statement certainly at odds with Banigan's obvious antagonism toward the Order at the plant. He excoriated the operatives for intentionally trying to spoil thousands of dollars of rubber stock. He cited the moral obligation of cutters to be a "natural guardian of that work until it is finished," a remarkable insight into a disappearing belief system of artisan culture and responsibilities that Banigan's tactics and the rubber industry slowly undermined. He praised the fifty loyalists who stayed behind to limit the loss, many now appearing on the Knights' boycott list. He finished the letter by promising to open his books to "any prominent, disinterested person, either a business or professional man" to inspect the operation's bottom line. Later in the walkout he offered the Knights of Labor $50,000 to prove he was not telling the truth about wage levels, and then proposed selling company stock to employees at $25 a share cheaper than speculators were paying. While some of this may have been bravado, Banigan was enough of a loose cannon on the local industrial scene to draw a pointed condemnation for threatening to disclose company ledgers. According to the editors of the *Providence Journal,* that suggestion would "hardly be accepted as an obligatory precedent." The mainstream press stigmatized Banigan for bypassing his employees about the news of a pay cut, although such criticism was seldom leveled at Yankee employers, who rarely provided any information about corporate decisions. The media then took Banigan to task for a progressive open book policy that no other local industrialist would introduce or dare implement, especially during a work stoppage.[21]

The Knights answered in a well-written letter. They repeated their grievances for striking: reduced wages, arrogant overseers, and the dismissal of Joseph Hines. An unnamed leader reinforced the union's resolve and reflected on the idealism of the Great Upheaval: "We are not in want of money by any means, as we can have all the aid we desire from the Knights of Labor in other parts of the country, and which is one of the strongest workingmen's organization in the country, stands by its members firmer, and conducts its business in the best possible manner." He reiterated that "the organization but recently had a bitter fight with the Wabash railroad, and had the pleasure of breaking the back of Jay Gould. The strike now in progress here will be fought to the bitter end; nothing will be left undone, as we are bound to have our rights. We can break any concern in the United States, and snap our fingers at any of the kings." The victory over Gould, the quintessential bad guy of the period—even in the eyes of other capitalists—gave an aura of invincibility to the thousands of workers who now joined the Order nationally and in Millville, but over time almost bankrupted the Knights. The larger issues for the rubber industry came in a jeremiad from Charles W. Hoadley, head of LA 3862 at the Para Rubber Company in South Framingham, which had also struck its employer and fashioned a progressive settlement. He accused Banigan, in another widely reprinted letter to the editor, of being the "head of the rubber trade of the United States, and his word is law in his own sphere—a magnate of most unlimited means." He went on to cite a half dozen other rubber firms that lockstepped in line immediately after the Woonsocket Rubber Company decreased wages "until all the operatives in the country had felt the hand of the oppressor." The characterization of Banigan's influence was not at all unfounded, as exemplified by the rubber cartel courting him so assiduously the following decade.[22]

The walkout temporarily remained a war of words, picnics, and peaceful rallies. Local newspapers remarked favorably on the absence of public drinking, an anomaly in Rhode Island, which provided the usual left-handed compliment to the Irish and the temperance movement. The *Woonsocket Evening Reporter* commented that "The rubber works strike is of such a quiet and undemonstrative nature that its existence is hardly perceptible, except in the noiseless and motionless machinery." The Knights obviously enforced good behavior. They also shunned a handful of employees still at work. "They are placed under a social and business ban," the editors of the *Providence Journal* scolded. "They can neither be spoken to nor traded with." Their names appeared in local newspapers and in circulars distributed to neighborhood businesses threatening a boycott against shopkeepers who traded with the enemy. One scab was derided as both a union and Irish turncoat who stole money from a local skirmishing club. A letter by a striker explained the older origin of their action: "Boycotting in Ireland received the blessing, the approbation, and sanction of priests and prelates. . . . It is as necessary in Millville as it is in Ireland, or Poland, or Russia." Another opinion expressed fear of the new weapon: "Is this community to be kept constantly in fear and trembling because of acts of violence, with the talons of boycotting to grope about and clutch by the throat any and all." Later in the strike Hines warned Banigan of just such a possibility: "We will use Capt. Boycott in your market," and then listed several of many out-of-state firms hurt by the refusal of Knights and their families to patronize offending culprits. The Millville Order officially endorsed social shunning as "far more effective than club or pistol."[23]

Violence

Two incidents of disorder shattered the uneasy peace. On 28 July, ten vigilantes "escorted" a strikebreaker out of town with a volley of shots over his head. Two nights later approximately sixty men, marching in columns of four and "acting as if under military discipline" attacked the Millville house of another turncoat who helped save Banigan's stock on the first day of the walkout and who now boarded scabs. Before exchanging bullets at 2:00 in the morning, the assailants woke and warned neighbors to take cover—a brazen, cocksure show of force to a community already solidly behind the strikers. Windows were shattered but Horace Childs, the target of the wrath and on the public boycott list, shot back and apparently wounded two invaders. The standoff took place just a few days after the annual statewide excursion of Clan-na-Gael, the Irish nationalist group that had a military component like the statewide Hibernian Rifles. These militant organizations evolved out of the failure of local Irish Americans to form an ethnic brigade at the time of the Civil War. Woonsocket and Millville boasted several affiliates. The Knights disclaimed any responsibility for the attacks, but the *Providence Journal* reported a recrudescence of Molly Maguire activity in the coal fields of Pennsylvania and, by implication, among the Irish rubber workers: "Millville was never in such excitement, and large numbers are armed." Industrialists, political figures, and the press usually portrayed the Order as an updated front for the Mollies, a group that periodically meted out Old World peasant justice against the British in Ireland and Yankee coal operators in the United States in the nineteenth century. Alan Pinkerton branded the Knights as an amalgamation of Irish secret societies in a framework of the leftwing Paris Commune. "In other words," according to historian Kevin Kenny, "the violence in which the Molly Maguires undoubtedly engaged was put to all sorts of uses by contemporaries, most effectively by those who were opposed to Irish immigrants and organized labor." Local authorities followed the national script. Selectmen from Blackstone, Massachusetts, which had jurisdiction in the village of Millville, stayed overnight; local sheriffs mobilized; and almost the entire force of state detectives came to town. The mayor of Worcester, Charles Reed, a manufacturer, sent twenty-five police officers to Millville, prompting a protest rally in anti-union Worcester by several thousand Knights.[24]

The mayhem apparently broke the logjam between Banigan and his employees. During the first week of August a committee of seven Knights led by Joseph McGee, who would hector Banigan for years to come as a Democratic town official in Woonsocket, met with the proprietor himself in Millville for three hours to discuss grievances. Both sides reported a courteous airing of complaints, although the usual confusion about who said what muddled any permanent understanding. However, they convened again on 8 August at Banigan's Providence office for a chilly four hour session after he had been tipped off, erroneously, that the union was about to capitulate. The Knights said his demeanor had changed since the earlier meeting and that he was interpreting the latest visit as a victory for his hard-line tactics. Hines wrote that "his usual bland style greeted the boys, but the twinkle of his cunning eye told them to look out for a storm." He summarized the negotiations as providing little agreement but more insight into Banigan's chameleon nature: "The committee could then readily understand the change artist of the Knight of St. Gregory." If Banigan had crossed the line with his errant but arrogant attitude toward his employees at the gathering, the mocking of Banigan's prestigious Catholic accolade took the operatives across the same terrain when they belittled his honor and, by implication, the Church.[25]

In the middle of August 1885, Banigan purchased one million pounds of raw rubber from Brazil at 51 cents a pound, the lowest price in the history of the industry. Even if he was unable to use it in Millville right away, he could employ some in the Woonsocket plant and sell the remainder to his competitors at a healthy profit and not worry about income during the walkout, because the seasonal deadline for orders was only a month away. The stockpile certainly helped his bargaining position. Banigan once traveled incognito to observe the harvesting of rubber at the headwaters of the Amazon River, probably one of the reasons why he understood the importation end of the business so well. He liked to tell the story that Brazilian businessmen warned him about a wily rubber producer in the States named Banigan![26]

Although the rubber king cringed at the public disdain heaped upon him by the Knights of Labor, he understood their growing power and studied the Order's rules and regulations. He sent a letter to Terence Powderly, General Master Workman of the Knights. He apparently included a copy of his public correspondence and added: "The men now on strike claim to be members of the organization of which you are an officer high in authority, and for which organization I have always entertained feelings of great respect, but the course which they have pursued has caused me to entertain serious doubts as to the validity of their claims." Although Banigan did not say so directly, perhaps letting the enclosure speak for itself, he questioned the actions of the Millville Knights in striking without first asking for arbitration as clearly required in the union's constitution—a process that Powderly steadfastly embraced despite its controversial nature in some labor quarters. Banigan studied the document, as he did with most things, and concluded that "you should take some action to protect the good name of your order in this vicinity."[27]

While Banigan waited for a reply from Powderly he publicly chided the LA: "If the rules laid down in your book had been brought to bear in this case it is safe to say our mill would have been running steadily up to the present time." Powderly apparently met with Banigan sometime toward the middle of September; he scribbled on Banigan's letter "answered in person at N.Y." The *Providence Morning Star,* a staunch supporter of the Knights and sympathetic to Irish immigrants, reported that "after stating his case, Mr. Powderly said if what he had said to him was so, he did not blame Mr. Banigan." Furthermore, Banigan allegedly replied he would rehire the entire workforce except for thirty-five or so individuals who had engaged in violent or outrageous behavior at the shootout but, if Powderly thought he should take the delinquents back, he would do that too. Powderly, according to the press report, advised against it! The General Master Workman never issued a clarification, lending credence to the interpretation that he probably washed his hands of the affair. Powderly probably had more in common with the upwardly mobile Banigan than with his rowdy rank and file membership. Carlton attacked Banigan publicly for trying to involve Powderly. The rubber king, without mentioning that he initiated the correspondence, said the labor leader "unexpectedly" dropped by the New York City office of the Woonsocket Rubber Company. Banigan pursued his advantage and claimed to have a letter from the national Order disowning the Millville strikers and labeling them a rabble. He was probably paraphrasing his conversation with Powderly and embellishing it to split the rank and file from the leadership, as no letter ever surfaced. Powderly never contradicted Banigan's remarks, and historical profiles of the General Master Workman point to his need for approbation from the elite, in this case Banigan, one of only a handful of Irish Catholics so prominently situated on the other side of the great personnel divide who usually practiced progressive labor relations. Powderly consistently advocated peaceful, educational methods to settle disputes. Two years later in a speech in Providence he declared: "I do not believe in them [strikes], and I have never ordered one, and I never will."[28]

Solidarity Forever

If Banigan were willing to raise the ante and spread the geographical boundaries of the debate, the Millville strikers were equal to the task. They had issued an earlier broadside entitled "An Appeal for Aid for Striking Rubber Workers" from Millville LA 3967, signed by Joseph Hines, the dismissed president, and Joseph McGee, the recording secretary and strike committee chairman. The authors outlined their grievances in a warning: "But to carry on a successful battle we must be equipped with the Sinews of War, and it is in order to furnish these requisite supplies that we make this appeal for assistance to our fellow-laborers in Massachusetts, and other sections of New England." The flyer got to the heart of the strike: solidarity to maintain their standard of living. "Victory for us, brothers, means victory for you, and if our banner falls in the conflict, surely your standard must come down along with ours."[29]

Although activity in the Knights of Labor rose dramatically with the stunning victory over Jay Gould in 1885, the Order failed to fund a crescendo of subsequent strikes. Special appeals for financial aid overwhelmed the Knights, despite a cordial reception at least initially by other workers in awakening industries. The Order directed the explosive energy of frustrated workers, anxious to repay less cordial employers than Banigan, into strikes, boycotts, and working-class political activity. Before these actions proliferated nationally, dues and contributions bankrolled early initiatives and often caught captains of industry and law enforcement agencies off guard temporarily. Operatives at the Para Rubber Company in South Framingham, now headed by Charles Hoadley, raised $400 from the rank and file for the Millville strikers. Another $200 came from local shopkeepers. Donations from businesses raised some hackles about the use of high-pressure tactics, but the Knights adroitly exploited the underlying tension between large companies and community-based, family operations. Banigan com-

AN APPEAL FOR AID
FOR STRIKING RUBBER WORKERS,

SANCTUARY OF LOCAL ASSEMBLY, 3967

MILLVILLE, MASS., AUG. 6, 1885.

To the Members of the Order of Knights of Labor, and especially those employed in Rubber Factories.

Our fight is theirs, and theirs is ours.

BROTHERS AND FRIENDS:—

On the 29th day of June the members of the above assembly came out on strike, and since that time have spared no pains in trying to form L. A. of the K. L. Their efforts have been successful, and have already formed two Locals, while another is being formed in Woonsocket, R. I. The first meeting between employer and employees took place Monday, August 3d.

A committee of the K. of L. waited upon Mr. Banigan, President of the Woonsocket Rubber Co., he (Mr. Banigan) having consented to meet them, although up to this time, he altogether ignored the Knights of Labor, and persistently and emphatically refused to recognize, or receive any individual or committee belonging to that organization.

However, a little time has wrought quite a change (for the better) on Mr. Banigan, as was evidenced last Monday when the gentleman received the Knights' Representatives in a most polite—yes, even a cordial manner.

At that interview, which lasted three hours, President Banigan offered us some concessions, which, on mature deliberation, we consider inadequate, and must as a consequence decline. We entertain most sanguine hopes, however, that a few weeks' more resistance, at the farthest, will bring us the objects we wish to attain.

But to carry on a successful battle we must be equipped with the Sinews of War, and it is in order to furnish these requisite supplies that we make this appeal for assistance to our fellow-laborers in Massachusetts, and other sections of New England. We deem it necessary to state here, that a former strike, which occurred in our shop last February, left many persons in a poor way to commence another campaign.

Therefore, fellow-toilers, we trust our appeal will not be in vain. We are determined to fight this battle out—ay, even as the illustrious departed Chief once said—"If it takes all summer." We ask not for privileges, we simply seek for justice. That justice we are bound to obtain, or else superior force will vanquish us. It rests with every wage-worker whom this circular shall reach whether the result for us shall be victory or defeat. Victory for us, brothers, means victory for you, and if our banner falls in the conflict, surely your standard must come down along with ours.

We have good soldiers of the rank and file. We have able and discreet commanders—yes, and we have a formidable opponent to encounter.

Will the outcome be a Waterloo for us or our adversary? That entirely depends upon our recourses. Help to furnish the sinews of war, and victory must certainly perch on our banners.

Fraternally yours,
JOSEPH F. HINES, M. W.,
JOSEPH J. M'GEE, R. S.

The local assembly of the Knights of Labor published a public appeal signed by its dynamic local officers. The strike, part of the Knights' Great Upheaval, sought the assistance of fellow workers. (The American Catholic History Research Center and University Archives, The Catholic University of America, Washington, DC.)

plained in a letter to Carlton: "Those people have been around begging in Woonsocket and Millville, to my knowledge, and the first persons they went to were liquor dealers," another of those ethnic insults among the Irish at different ends of the industrial world. Interestingly, another circular distributed in South Framingham forged a telling link between prosperous producers and small businesses:

> At present we are waging a battle with the Woonsocket Rubber company which promises to be of great import to this town as regards its future prosperity: for if low wages prevail there, look out for depreciated real estate, bankrupt merchants, and tenantless homes in South Framingham. So sure as the wage workers meet with defeat there, just as sure will you feel the oppression of the monopoly that controls the rubber market of the world.

Within seven years that prescient warning materialized in the formation of a new Gilded Age cartel, the United States Rubber Company, in 1892. Other donations came to Millville, and the Massachusetts district assembly made several outlays to strikers with families. The *Worcester Daily Times*, edited by a member of the Order, James Mellen, listed contributions on the front page, usually in small amounts. A fundraiser in Braintree, Massachusetts, raised $2,000 by selling $1 tickets after the strike.[30]

Banigan employed new tactics to pressure his operatives, many of whom had left town by September for work at other rubber companies. He apparently tried to influence his colleagues to blackball them, although there was no love lost between rival rubber magnates. Most competitors coveted the skilled workforce assembled at the Millville plant and some regularly sent agents to the area to hire Banigan's help even before the walkout. The Cambridge Rubber Company, the Para works, and startup Franklin Rubber plant—just a few miles away and a major stop on the Providence and Worcester Railroad from Woonsocket—were hiring his employees as fast as they arrived, especially more transient single males. On the other hand the Colchester, Connecticut, rubber plant, in which Banigan held a controlling interest, refused to take Woonsocket workers because of a blacklist. Furthermore Banigan had built a limited number of comfortable tenements in Millville in order partially to ease a severe housing shortage, one reason why so many operatives still lived in nearby Woonsocket. By midsummer he began to expel strikers, including Joseph Hines and his mother, from the compound known aptly as Banigan Village. At about the same time the American press, including the Woonsocket newspapers, reported on the eviction of tenants in County Kilkenny, Ireland, a convergence of seemingly similar events not likely to help Banigan's public image among local Irish immigrants, who cringed at any actions that smacked of displacement by authorities at home or abroad. Although the Banigans had departed Ireland voluntarily, the subsidized emigration to leave their homeland always smacked of forced exile among the dispossessed. The irony of the situation probably did not escape the local community or the strikers.[31]

Banigan needed either to accept orders for the rapidly approaching fall production deadline or close the mill for the season and rely on other sources of income. He claimed to have stockpiled 10,000 pairs of boots and sent a notice to jobbers to buy now because there were "no prospects of the operatives going to work." Even though he had the wherewithal to wait it out, he risked losing his crackerjack workforce for good, never to regain the competitive personnel edge that had put him at the top of the cutthroat rubber footwear business. Banigan threatened to recruit large numbers of permanent strikebreakers, probably as a stopgap pressure tactic before actually turning to that as a last gasp strategy. Earlier attempts at importing substitute workers a few at a time had failed, as strikers met all incoming trains and usually cajoled, threatened, or paid passage home for the putative replacements. The informal solidarity between operatives in the industry and the highly skilled nature of the bootmakers soured that scenario. In the second week of September Banigan announced that he was hiring three hundred scabs from Massachusetts and, as a kicker, they were mostly native-born Americans who qualified to vote, another ethnic slap at some of his disenfranchised immigrant countrymen in restrictive Rhode Island. He also stated that more than one hundred picketers stood ready to return to the mill immediately. As an inducement Banigan promised to pay weekly rather than biweekly wages, one of the great desires of workers in that era. With so much movement swirling around the strike in September, the *Woonsocket Evening Reporter* refused to print any more external communications about the increasingly publicized walkout because of the "large number of outsiders who seem to regard the issue as a sort of free fight, in which anyone can take a hand," a reflection of the larger battle between wealth and commonwealth in Gilded Age America.[32]

As the strikers had moved Banigan to the bargaining table earlier in the walkout with their violent offensive, Banigan now nudged his employees to another meeting after two months on the line, with the threat of a legion of strikebreakers. On September 14 the two forces collided again. The executive board of the Massachusetts Knights of Labor, DA 30, traveled to Millville to confer with Banigan, who had pledged to set aside an entire day to settle the walkout. Joseph Hines greeted the Bay State delegates, all of whom had national reputations: Albert Carlton, George McNeill, Charles Litchman, and Frank Foster. Several hundred strikers also welcomed the union officials while ignoring Banigan, who got off the same train. Police were out in force. Reporters hailed from New England and New York. Rumors had it that a settlement was forthcoming. Instead, after meeting for several hours, the Knights balked at any agreement upon learning that sixty scabs were on their way from a Boston employment agency. They walked out when the rubber king refused to halt the invasion, accusing him of "bringing up reinforcements under a flag of truce." Carlton threatened to ratchet up the situation by repeating an earlier threat: "We feel at liberty to invoke the aid of every brother workman that is employed in other factories that you represent, and also every Knight of Labor man in the country, for the purpose of boycotting all goods bearing the trade marks of your factories." With so many working-class miners wearing his boots, and homemakers using his clothes wringers, Banigan had to take this warning seriously.[33]

Father Michael McCabe

The strike at Millville severely crimped the region's economy as payroll losses mounted to over $100,000. As accusations and recriminations ricocheted from one party in the dispute to the other, Father Michael McCabe, the Irish-born pastor at Woonsocket's St. Charles Church, stepped into the fray to stop the internecine warfare that was tearing the Irish community apart on both sides of the state boundary. McCabe was born in County Leitram in 1826. He attended a seminary in Baltimore and was ordained in Providence in 1854. Assigned to

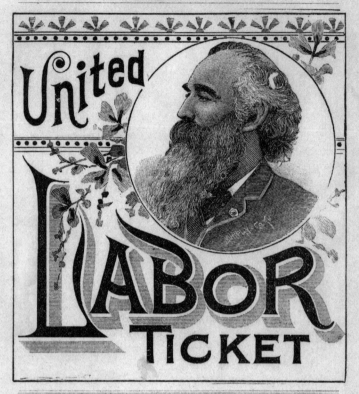

George McNeill, one of the most prestigious leaders of the Knights, visited the Millville strikers. He authored one of the era's seminal monographs about the "labor problem," and later ran unsuccessfully for mayor of Boston in 1886, a year after the Banigan walkout. (Author's collection)

Father Michael McCabe, the Irish-born prelate, who mediated the strike between Banigan and his workers. McCabe led St. Charles parish in Woonsocket and served as the Church's vicar general of the Diocese of Providence. (From James Smyth's A History of the Catholic Church in Woonsocket, *1903)*

Woonsocket a year later, he implemented a successful long-term plan to erect a fashionable $100,000 church. He established the city's first parochial schools, brought the Sisters of Mercy to town, and was an articulate spokesman for the Catholic community. He became vicar general for the diocese in 1879 despite spending most of his career outside the religious nerve center of Providence. His emphasis on a disciplined, law-abiding, temperate Irish garnered the attention and praise of the Yankee establishment as well, despite his simultaneous support for Irish nationalism and Irish-American activity. He visited Ireland at least five times while in Woonsocket and donated $25 to the nationalist Charles Parnell Defense Fund.[34]

During the original strike in February, McCabe had entered the conflict as an erstwhile mediator, visiting another rubber shop in order to compare conditions. Six weeks after the initial conflict Father McCabe celebrated the final mortgage payment on St. Charles. Before the predominantly Irish-American congregation, he handed over the deed to the parish's trustee John Holt, the superintendent of the rubber company, Banigan's father-in-law, and a Catholic philanthropist in his own right, especially in Woonsocket. On this joyous occasion the *Providence Journal* weighed in with an unusual accolade to the clergyman: "He is to be congratulated, not only by his Catholic superiors and priests, but by all good Protestants." The newspaper's wishes also reflected on his free enterprise orientation and, more important, his achievements in helping to tame and transform the wild Irish into productive citizens and workers, although it had been rumored during the Civil War that McCabe had counseled Irishmen not to fight. Furthermore, some Irish parishioners resented his involvement in a quarrel with the Ancient Order of Hibernians, a fraternal and cultural society, in 1878. At the time McCabe had called AOH members Molly Maguires, a charge emanating out of the show trials in Pennsylvania during the same period.[35]

Now, in September, McCabe took the unusual step of announcing at Sunday mass a special meeting for striking parishioners the following evening in the church basement. However, in his Sabbath sermon he preached against violence toward scabs, who, he said, had the right to work for nothing if they so desired. Despite a Knights' gathering elsewhere, seventy-five came to hear the priest. McCabe familiarized himself with the complex issues and sought to put out the fire, but initially only fanned the flames. He shared a proposal from Banigan that would match wages at the Boston Rubber Shoe Company in Malden, a direct competitor. McCabe also stated that some supervisors actually supported strikers' demands, perhaps trying to heal the rift between shop floor management (like the fatherly John Holt) and point-of-production workers.[36]

Despite McCabe's conciliatory approach, he was interrupted several times by strikers unwilling to trust any traditional authority. McCabe, not surprisingly, was an unapologetic friend of Banigan. In other situations he praised the rubber king's philanthropy, such as the opening of the Banigan-financed Home for the Aged Poor in Pawtucket a year earlier. Near the end of the meeting, Joseph McGee, the local union's secretary, charged that McCabe had been bought off. The priest replied that his only concern was the welfare of his parishioners, and added that the thirty-five workers Banigan refused to re-hire would be facing legal action for the shooting incident in July if not for his intervention on their behalf. McGee then informed the priest that the strike had gone too far for concessions and that, at an earlier union meeting, members had voted unanimously not to attend any further sessions with him. Stunned by the accusations, McCabe left but apparently resolved not to abandon the cause of mediation.[37]

The priest immediately wrote a letter to the editor of the *Woonsocket Evening Reporter* that appeared beneath an article describing the bitter meeting. "I am no delegate to the Rubber company or any other," he rebuked. "Let them not credit any crank who may tell them I am bought out to work against their interests. The Rubber company has not money enough to make such a purchase." The *Providence Journal,* sensing an opportunity to renew its attacks on Irish-American workers that stretched back a half century, editorialized "that when a thousand Irishmen neglect the fraternal approach of a man like Father McCabe, they are likely to suffer the penalty of rashness and improvidence." The vicar general would later admit his humiliation at a consecration in Hartford, Connecticut, where one hundred priests from the region assembled. "Each and all," he said, "questioned him regarding the Millville strike, and all were greatly surprised to find that a Catholic congregation could be found that would not listen to the voice of their pastor." Such occasional rebellions, here and abroad, forced the Catholic Church to tighten discipline and squeeze out unsanctioned cultural and folk practices practiced by different ethnic groups. Although McCabe may have exaggerated the situation, the Church truly feared any independent action through the auspices of the labor movement that might elevate economic, material concerns over spiritual ones, especially if such activity led to questioning of ecclesiastical authority.[38]

McCabe, who decided to speak with the strikers again, attracted only a handful, an indication of the Knights' growing influence and power in the community. Usually the Catholic Church was the agency banning others; in a role reversal, the Knights interdicted McCabe. At the next mass the pastor tried conciliation once again by emphasizing a balance of interests between labor and capital. Then the pastor played his most influential trump card, the always sensitive ethnic question: "Are there not many men who came here from Ireland who received good wages, who, if they did not get work in the rubber works of Woonsocket would be obliged to go to the coal mines of Pennsylvania and work for starvation wages; or in the building of railroads amid the snows of winter in the far west?" Ending his sermon, he reiterated that he was not beholden to Joseph Banigan and to prove it said he had visited Malden again during the week to investigate prices and wages there. He also stopped at the Knights of Labor headquarters in Boston and discussed the possibility of reducing the number of blacklisted employees. McCabe, more than most key players in the conflict, understood the potential damage to the reputation of the Irish-Catholic community in the ongoing fraternal quarrel. The Church's flock, plebians and philanthropic contributors alike, was jeopardizing years of hard work by McCabe and others. Public arguments between the Knights of Labor and the Knight of St. Gregory embarrassingly paralleled the derogatory language and stereotypes used by their enemies.[39]

Near the end of the walkout, Banigan closed the still-operating Woonsocket plant temporarily because of overstock, but the *Providence Journal* wrote that he feared an imminent walkout there. The striking bootmakers probably instigated the first rubber workers LA ever in the Knights in 1882, at the more automated Woonsocket shoe firm. When boot production shifted to Millville, so did the artisans, leaving behind a predominantly female staff in an increasingly mechanized shoe factory. Not surprisingly, the original LA was replaced by a more representative "ladies assembly" chartered in Rhode Island during the middle of the walkout. That union, which reflected the changing gender base of the workforce there, lasted until 1894 and demonstrated once again the Knights' understanding of the need to organize skilled and unskilled, male and female, even if that meant at times retreating into segregated LAs. The women at the home factory in Woonsocket faced great pressure to join the walkout but may have vacillated without the presence of almost irreplaceable, skilled male bootmakers and cutters, as well as their fear of unskilled substitutes in such an automated facility.[40]

Settlement

With the continuing animosity and the possibility of the Woonsocket plant going out, Banigan met again with Carlton and the Knights' committee, minimizing the hiring of scabs by calling it part of a planned enlargement of the workforce. Both sides finally agreed to accept a compromise wage scale, and Banigan promised to consider rehiring some of the thirty-five "obnoxious men." Although he refused to reveal the names on the list, most of them knew they were marked and took other jobs. Banigan further pledged to introduce weekly payments for several months, and perhaps permanently, in order to help workers get back on their feet financially, another workplace concession not countenanced by his fellow Yankee manufacturers. On the other hand, the hometown *Woonsocket Evening Reporter* lauded Banigan's position: "the first large corporation in the state to grant this boon to its workmen."[41]

The leadership of the Boston Knights was ready for a settlement after an exhausting battle that had tested the limits of everyone's patience. The executive committee of District Assembly 30, acting unilaterally as an arbitration panel, issued a directive to the Millville strikers on 29 September: "the board, considering the duties its members owe to the members of the order and to the community at large, and desiring to strengthen their system of arbitration for the settlement of difficulties, *has ordered the men to resume work*." The Boston leadership, with a wider perspective on the Great Upheaval than the rank and file, understood the inability of the Knights to withstand such an explosion in membership in tumultuous situations that threatened to overwhelm the Order's very existence, almost as much as any employer hostility. With so many ongoing strikes, DA 30 attempted to settle this intractable episode, once and for all, and move on to other situations. In retrospect, the whole affair might have been prevented by earlier intervention, but the demands of the Great Upheaval stretched the resources and personnel of the Knights too thinly. Carlton and McNeil, who authored the decision and tried to placate both parties, agreed with Banigan that the local assembly broke national rules by not seeking intervention. The board softened the interpretation and, in some ways contradicted it, by saying the grievances nevertheless justified a walkout. With an agreement on a wage scale now tied to the Boston Rubber Shoe Company, Banigan's biggest challenger, the Millville strikers concurred unanimously with the decree as the number of strikebreakers swelled ominously. The Boston Knights also promised to take the thirty-five untouchables under their financial wing, knowing full well that Banigan would probably hold the line in regard to them. The *Providence Journal* applauded the boss's decision against the delinquents: "It is a question of public order, of public safety, of the maintenance of law and order. Herein the company is squarely on the side of the peace and security of the community. It represents protection to the individual however weak. It refuses to offer a premium to mob law, and the torch of the midnight avenger of a fancied wrong." Whether blackballed or not, the leading spokesmen for the Millville Knights took other jobs as well: Joseph McGee became a city councilor in Woonsocket, Joseph Hines opened a grocery store in Millville.[42]

After thirteen weeks on strike the bootmakers went back to work on 5 October, 1885. On opening day approximately 334 employees boarded railroad commuter cars, "about the whole number of residents of Woonsocket employed in the Millville works." The *Providence Journal* prematurely wrote that the plant "will once more assume its industrious aspect of old, and harmony and peace will reign supreme." On the third day, employees received individual contracts to sign. There were twenty-one rules, boilerplate language typical of factory employment in that era and similar to regulations before the strike. However, the last clause read, "Regular attendance at some place of public worship, and a proper observance of the Sabbath, will be expected of every person employed." This new, unanticipated requirement triggered another walkout including almost every one of the original strikers as well as a few of the 235 scabs and various new hires. Some simply refused to sign the document as contrary to Knights' policy, while others opposed any faith-based regulations. Although similar ordinances frequently appeared in Yankee establishments, perhaps the Irish-Catholic workforce which seldom dominated local factories, took it as a personal insult in this instance from a boss who knew better. These immigrants and their extended families had the reputation of seldom missing religious services and recoiled at a directive that suggested otherwise. Banigan may simply have included the language as a final taunt to his unappreciative help.[43]

In the land of Roger Williams, where secular authority gingerly avoided any perceived encroachment on religious freedom, "Banigan's commandment" was roundly condemned by just about everyone. The *Providence Journal* opposed it: "for the great majority of the operatives are understood to be Roman Catholics, who, as such, are noted for their regular attendance at mass. . . . Now, however conscientious the motive of said President [Banigan], it is not in accordance with the spirit of American institutions that any man should dictate the private conduct of the individual in religious matters, and control him when not in his employment." The *Providence Morning Star* angrily stated: "He ought to be the last man to be a party to any act which might ever seem like interference with the religious thought and actions of others. . . . [He] belongs to a race which, until quite a recent date, labored under severe disability, imposed by British law, upon Irish Roman Catholics." The *Worcester Daily Telegram* wrote that "Banigan of Millville is an Irish Catholic, but that does not save the overbearing fellow from the execrations of wage-workers of the same race and creed."[44]

The *Woonsocket Evening Reporter* published many related letters. One complained that, "They would try to shove a man into heaven by brute force." Another missive made the ubiquitous connection to the fatherland: "God knows we Irishmen have suffered persecution and injustice long enough in the old land, and it is truly lamentable and overbearing to have to suffer the same here in free America at the hands of our own countrymen." Still a third letter remarked that before the walkout, when a worker had a complaint, the company refrain was "go to hell or Framingham"; now with the new contract language the song went "come to heaven and the rubberworks." Banigan's reputation, like that of the strikers, ebbed and flowed with the undulations of the walkout. This firestorm sullied him, although several other communications, in-

cluding a few from a Protestant clergyman, interpreted his pronouncement differently: "As a manufacturer he sees, what all must see if they
will look, that men who know the Sabbath and attend the House of
God are, as a rule, more trustworthy than those that do not." Some of
the skilled workforce of cutters and bootmakers had had enough and
went to the nearby Franklin rubber facility where the proprietor had
an open invitation to hire Millville rubber workers regardless of their
membership in the Knights or whether they went to church.[45]

The editorial salvos aimed at the boss provided some comfort to the
strikers, who saw a confirmation of their characterization of Banigan
as authoritarian. But the Boston Order must have tired of the constant warfare and yet another walkout. Father McCabe revealed in a
public letter that: "The executive committee of the [Mass.] Knights of
Labor, according to their usually reasonable actions, have withdrawn
their objections to their members signing the [church attendance]
rules lately adopted at the Millville rubber works." The accommodat-
ing DA 30 authorized Father McCabe to oversee the process as long
as Banigan adhered to the accepted wage scale and rehiring of strikers. The Boston leadership must have wondered if anything could
solve the seemingly unending imbroglio. The bootmakers quickly decided to go back. Most of them were regular churchgoers and felt the
section was irrelevant, petty, and unenforceable anyway. Banigan,
however, had overstepped his authority and misread the temper of the
situation. The strikers nailed him as handily as a leather sole over a
rubber lining.[46]

As the rubber workers returned to the plant once more, Father Mc-
Cabe then shocked a community already used to daily surprises. The
final wage settlement that ended the long strike was consummated by
the priest and John Holt while Banigan apparently visited Ireland.
The rubber king may have been as stymied and frustrated as the
Knights and felt that the two less divisive figures might be able to fashion a settlement in his absence. When Banigan returned right after
the pay agreement and during the religious protest, he apparently
scolded McCabe and Holt for an unsanctioned compromise, although
he more than likely had empowered them to forge a reasonable ad-
justment. Father McCabe probably felt that Holt, whom he called "an
honorable and truthful man," was authority enough for the accord.
However, in a stunning letter to the editor, the pastor lashed out when

Banigan appeared ready to scuttle the contract: "I made promises to the Millville bootmakers," he wrote in a widely reprinted piece to the *Woonsocket Evening Reporter*, "which Mr. Banigan will not sustain. He tells me I have no right to make promises, as I am not running the works. . . . If I were running the works I could have settled the matter long since. . . . I only hope Mr. Banigan has done wisely for himself and the community in refusing to endorse the settlement which we thought was made, and which would have been satisfactory to all." The clergyman's upbraiding isolated Banigan from his own Irish-Catholic constituency, while the state's power structure enjoyed another Irish tiff. The Church finally abandoned its role as neutral mediator when the vicar general chastised the Knight of St. Gregory. Banigan back-tracked immediately. He claimed he initially misunderstood the com-promise wage structure but now endorsed it upon further study. Fi-nally, both sides consented to the original terms and probably agreed to informally ignore the worship fiat. Father McCabe arguably felt he reinstituted church authority without any assistance from Banigan.[47]

The Irish Question

At the conclusion of the strike, the community offered different in-terpretations. The *Woonsocket Patriot* passed it off as a Hibernian quar-rel: "We have a profound respect for the Irishmen; they are good fighters, but often do not exhibit good judgment in fighting," as in the recent strike which shook "the confidence of Americans in their ability for self government," a not so subtle slap at Home Rule in Ireland. A letter answered the charges: "There was nothing Irish in the strike." The anonymous writer stated that the participants acted as a class: "they have heads to plan and hands to execute. They are no drones in the national hive . . . a few hundred rubber workers, Irishmen, if you will, dissatisfied with their business relations, went on a strike. A matter of no importance or significance except to the parties concerned."[48]

In a symbiotic embrace, the Knights of Labor flourished as the Irish came of age in Rhode Island, Massachusetts, and the United States. They were no longer refugees from the Potato Famine. Some families had reached third-generation status. The Order characterized the Millville strikers: "every man is himself a leader," a long way from ear-lier Irish immigrants in Providence so illiterate they signed their names on payroll stubs with an X. The organizing events at Woonsocket and Millville in 1885, as well as several generations of involvement in fraternal, religious, and nationalistic groups, elevated the strikers from alien Irish into experienced American citizens. The labor historian David Montgomery pointed out that in this era: "The formation of a trade union consciousness involved an important step away from purely ethnic loyalties and traditions."[49]

Banigan, a Potato Famine refugee, tried to act as an old-fashioned protector of his people until doing so threatened his bottom line. By the mid-1880s, however, skilled cutters and bootmakers were no longer young, awkward greenhorns or the wild Irish, but mature craftsmen in their own right. They were Americans and Irish Americans and they chafed at the patronizing tone of a few industrial benefactors as well as the father-child dichotomy of the Catholic Church. In some ways the strike was only, as suggested by one letter-writer, a shooting star in the labor-management constellation. On the other hand, the rubber workers bit the aristocratic hand that fed them and also forced the Church, as represented by Vicar General McCabe, to take their side unequivocally. These workers stood up to take partial credit for the flourishing rubber industry and the piercing spires of Catholic churches they helped subsidize. The bonds of deference had been loosened but not severed. Time and other crises would reunite the current adversaries. The estrangement, while not permanent, signaled a coming of age of this immigrant community. The collective step forward at this moment probably trumped—for the first time—the long journey of Banigan himself. The rank and file took a leap toward empowerment and independence in this event, but not before they walked in the rubber king's giant footprints while stepping ahead on their own.[50]

Joseph Banigan, the Knight of St. Gregory, must have felt very lonely during the 1885 strikes. Although the conflict ended in a compromise, along the way his employees managed to publicly tarnish the image of the rubber king among his most partisan supporters: the predominantly Irish workers themselves, his beloved Catholic Church, and a few enlightened elements of the Yankee aristocracy who increasingly though reluctantly had come to view him as one of their own fraternity. The Knights of Labor, in a rare historical moment, empowered and emboldened the rank and file rubber operatives to do the unthinkable and launch an attack on one of their own kind. Despite Banigan's honest bravado, he could no more overcome the solidarity of the Knights in Woonsocket and Millville than Evelyn Shirley and Steuart Trench could conquer the Monaghan peasants. Power shifted for a few short years in both instances, and the transfer of influence rattled the status quo. The open assault on Banigan's character from inside the Irish circle must have inflicted a stinging wound.

The individual Knight and the collective Knights threw the weight of their life experience into the fray. Banigan held no monopoly on being a Famine refugee as he did in the wringing industry or later in the rubber cartel. The author of his obituary in the *Providence Visitor* wrote that "he was reared in a school of adversity."[51] If he had ascended above his compatriots, he still stood on their stout shoulders whether he admitted it or not. Both management and labor revealed years of toil and sacrifice. Along the path of life all had suffered in some way. Each side drew upon a wellspring of determination that surprised one another. Employer and employee unexpectedly met their match and moved on.

TRAGEDY, PHILANTHROPY,
AND LACE CURTAIN

· ·

> *"He [Joseph Banigan] makes hair-breadth escapes."*
> —Woonsocket Patriot, *25 April 1873*

> *"The extent of the charities of the Church cannot fail*
> *to impress the Protestants as well as the Catholics."*
> —Providence Visitor, *31 May 1884*

Joseph Banigan engaged the unified might of the Knights of Labor with the characteristic vigor that marked any challenge he faced. Earlier in his life he had confronted other crises that tested his courage and forced him to consider his own mortality: to wonder whether he had inherited the proverbial "luck of the Irish." Like the nation's Puritan founders, he must have asked himself, in the midst of several tragedies, whether he was a visible saint or just an affluent sinner. Two catastrophes prior to the strikes stand out—experiences that must have terrified even one who had fled the Famine. Banigan came away from these with his charitable nature reinforced and his personal resolve strengthened.

Before these nightmare events, he suffered a personal loss. On 21 December 1867, a year after he moved to Woonsocket, his father, Bernard, died at age 68. Identified as a laborer in life and death, as were most of the Irish in Providence, the father of nine children had looked deep into the abyss of the Potato Famine and led his family to Scotland, back to Ireland, and finally to the United States. The miniscule rent he paid on the Shirley Estate placed him at the bottom of the food chain, literally and figuratively. He was a cottier who worked the margins of a potato field and the fringes of a destitute society. His will power to persevere was not unique, as attested to by so many Irish emigrants who eked out a life somewhere other than their homeland after journeys that had seemed damned from the start. Bernard Banigan, born in the physical and cultural barracks of the English ascendancy, as confining as any prison, broke his chains under the worst duress. Still, as with so many Irish immigrants heroic in their search for a safe haven half a world away from home, no obituary graced the pages of any Rhode Island newspaper when he died. To be sure, Joseph Banigan must have held a dignified funeral for his father and, many years later, he built St. Bernard's mortuary and chapel in his honor. He reburied him and other family members there as an individual and household tribute. An account of Irish street pavers in Providence during the Civil War unwittingly provided a collective obituary for all of these common laborers by describing them, while they worked with picks and equipment, as a "battalion of Melesians, who beat the dead march of Saul in rhythmical harmony."[1]

Several years after his father's passing, Banigan faced another family crisis when his wife Margaret died at age 27 on 4 April 1871. She was initially buried at St. Patrick's Cemetery in Providence alongside his father. The mother of their four children was born in Manchester, England, and came to Rhode Island with her parents in 1851. She and Banigan probably met through her father, John Holt, when they toiled together at the Providence Rubber Company ever so briefly. The courtship must have been easy to arrange, as the families lived a few blocks from each other. John Holt probably blessed the relationship, for he stayed close to Banigan for the rest of his life. Unfortunately, Margaret Banigan, like Bernard Banigan, garnered no known obituary, only the obligatory mention in the city's death record. Joseph Banigan would later reinter her body in St. Bernard's mortuary very close to the Holt family burial plot in St. Francis Cemetery in Pawtucket. As with most events at this time in his life, we have no direct knowledge of his feelings. These deaths must have stoked memories of the indiscriminate vagaries of the Great Hunger in Ireland that cut down relatives, neighbors, and friends regardless of age and condition, without warning.[2]

The *Metis*

If Banigan found comfort through his labors, he did so with peripatetic energy. He still personally orchestrated the search for customers as general agent of the Woonsocket Rubber Company and several affiliated concerns. He scoured potential marketplaces for his footwear and continuously traveled in and out of Providence by sea and rail. If the deaths of his father and especially his young wife had not already tested Banigan's faith, two disastrous events in 1872 and 1873 must have done so. On 29 August 1872, he boarded the steamer *Metis* in New York City, probably on a return trip from a business venture. The modern propeller ship measured two hundred feet long and carried passengers and freight. The vessel plied Long Island Sound and was part of the ill-fated Sprague commercial empire in Rhode Island. The ship, with Banigan aboard, left New York at 5:00 p.m. Although the passenger list officially numbered just over one hundred, the staterooms and the deck bulged with extra travelers, their children, and a crew of forty-five. Cargo stuffed every available spot. The water proved turbulent as usual in this agitated Atlantic shipping lane.[3]

Just before 4:00 the following morning, the *Metis* struck a schooner, almost imperceptibly according to witnesses, off the coast of Watch Hill in Westerly, Rhode Island. Although the boat had life-saving equipment, the officers seemed unprepared to deal with the crisis. Many of the passengers, ejected into the ocean as the craft sank, flailed in the water and perished. Banigan managed to survive. Because of the ensuing chaos and loss of life, the number of dead and survivors remained a matter of speculation, but the count of those who drowned in the wreck was about fifty. Bodies surfaced along the shoreline for several weeks after the accident. Only the *Woonsocket Patriot* carried any significant notice about its local prominent citizen: "Mr. Joseph Banigan, agent of the Woonsocket Rubber Company, was a passenger on the steamer "Metis," lost on Long Island Sound last week. He saved himself by the use of a life-preserver, aided by a cork matrass [*sic*] and although some seven hours in the water, made a miraculous escape."[4]

Banigan, except during the 1885 strikes, seemed private by nature. Perhaps at this early juncture in his life, before his star had risen, his ordeal or even the opportunity to discuss it, never presented itself. Later in life, however, he mentioned the tragedy in detail. He assisted a number of fellow passengers by helping them to secure and attach life preservers. Once in the ocean he rescued a woman struggling to stay afloat. He held her head above water until they grasped a mattress; it was several hours before another boat finally assisted them. Banigan kept a model of the *Metis* in his private office.[5]

Richmond Junction

There was another way to reach Providence from New York City that avoided some of the rough water the *Metis* and her sister ships traveled. A passenger could take a steamboat to Stonington, Connecticut, and connect with a waiting train to Providence directly from the wharf. The route, established in 1837, belonged to the New York, Providence, and Boston Railroad. Nicknamed the Stonington line, these cars carried both freight and passengers, just as the ships did. On 18 April 1873, Joseph Banigan boarded a vessel in New York and switched to the train in Stonington about 2:45 the following morning. The eight-wheeled-engine dragged three loaded freight cars and another five passenger coaches holding one hundred patrons. The front passenger car, a second-class affair known as the emigrant coach, usually carried newcomers from New York at a very inexpensive rate, with accommodations that matched the discounted cost.[6]

The train made a scheduled stop at Westerly as soon as it crossed the state line into Rhode Island from Connecticut. On this day, running a bit late, the engineer argued with his counterpart on the local express about who should go first. The operator of Banigan's carrier won the debate but in a pyrrhic victory. The engineer pushed the throttle to forty-five miles per hour. About nine miles later the train approached the Richmond Switch or Wood River Junction in rural southern Rhode Island. Just beyond the farming village there, the railroad usually rumbled over a twenty-foot long span. Unknown to the trainmen, a flash flood from a broken dam had undercut the granite bridge abutments leaving a skeleton without underpinnings; the rails were holding themselves up. The approaching train snapped the weakened structure and shot across the chasm, smashing into the bank on the other side. Coaches splintered and caught fire from the inside stoves. Passengers burned to death or drowned in the swollen, though normally shallow and placid, waters. As in the *Metis* disaster, bodies and limbs showed up in the river for several weeks. Eleven officially died. Bret Harte, the famous short story writer, authored a poem about the tragedy. Banigan, one of the passengers, survived the crash without injury, prompting the hometown newspaper to proclaim, "He makes hair-breadth escapes."[7]

If Banigan ever doubted his faith when his father and wife died, the escape from such horrific accidents, less than eight months apart, must have affected him stunningly. Religious beliefs and incipient philanthropy already characterized his life, but when Banigan emerged from the carnage of the accidents he must have sensed some form of divine intervention if only to warn him to balance the corporal with the ethereal, and not to forget or fail to memorialize those from the potato patch or the family roots in County Monaghan. He regrouped around his faith, the one institution that unequivocally represented his former homeland and was tenderly reproduced in the new. He probably reached deep into Catholicism for explanation and guidance, but whatever the specifics of his spiritual quest, he had survived once again and seemed to expand and prosper almost in order to perform charitable endeavors. His hard work and business perspicacity secured ever greater financial rewards. As the money eventually poured into his coffers, a significant portion of the treasure went to the Catholic Church, although some funds went to other religious and secular causes.

Banigan refused to flaunt his philanthropy (although not his wealth), reflecting a certain shyness in this field and perhaps fearing that too much generosity only increased the incessant pleas for assistance. On the other hand, the Catholic bureaucracy gladly replaced the Protestant work ethic with a Catholic one that called on the faithful to balance commonwealth and wealth. The Church wanted its models and financial saints to set a standard, even if those less fortunate had to strain even to approach the lofty heights set by a Banigan. Visible saints, abhorred by the Puritans, who distrusted the trappings of good works and financial success, eventually gave way to a free enterprise, Protestant work ethic. Banigan's good fortune could grace the altar without fear of compromise or contradiction in Catholicism as well.[8]

Banigan acquired a spiritual guide in Rhode Island who escorted him along the road to Irish-Catholic philanthropy. Although he never had an equal partner or associate in the rubber footwear business, he found a perfect match in the local clerical ranks who matched his corporate vision within a religious panorama. They formed, perhaps, the greatest tandem in the history of the state's Catholic Church. Bishop Matthew Harkins made life both easier and more difficult for Joseph Banigan. He strained the rubber king's fortune but may have fortified his zest for even greater abundance to sustain his charity.

Bishop Matthew Harkins

Matthew Harkins was born in 1845 in Boston, the son of Irish immigrants. He studied in France and became a Catholic priest in 1869. He served at several parishes in the Boston area beginning in 1870 and earned a sterling reputation. Pope Leo XIII, who knighted Banigan in 1885, appointed Harkins the second Bishop of the Diocese of Providence in 1887. He held office until his death in 1921, the longest tenure of any prelate in Rhode Island history. The new leader doubted his abilities every inch of the way, unlike his key ally Joseph Banigan, who never doubted much of anything in the world of business. They complemented each other masterfully. Harkins, like Banigan, refused to tolerate incompetence in any form. On the other hand, the open-minded businessman who ruthlessly pursued commercial interests, revealed a greater toleration as he aged, sometimes even in corporate affairs. He still bristled on occasion, as in the strike of 1885, when his own kind rebelled against his benevolent dictatorship, or when Yankee competitors challenged his leadership in the rubber industry.[9]

Harkins, on the other hand, disliked the mixing of different religions and the socializing of their respective followers, although he too seemed to soften on this issue over the years. Irish Catholics had suffered such an incessant, sustained bombardment by the time Harkins took office that a defensive posture flirted with outright paranoia, and not without justification. He recoiled from ecumenical services or even public ceremonies where representatives from other religions might address Catholics, especially school children. He harangued Catholic politicians or other priests for any incursions, however harmless, across the great religious divide. A half-century of virulent anti-Catholic rhetoric and discrimination in Rhode Island instilled caution in any local "papist" in the "Fortress Catholicism." Harkins, however, possessed the wherewithal to distinguish a true and honest alliance from a sham publicity stunt. After all, his predecessor, Bishop Thomas Hendricken, in worse circumstances, endorsed religious integration within the Knights of Labor that included "mingling" of every imaginable type in a diverse workforce. Bishop Harkins missed some great opportunities to foster toleration and acceptance of Catholicism, yet to criticize someone so promethean in so many other areas seems cruel. He was able to soar so high that, on those few occasions when he fell to earth, being a part of the temper of the times made him seem pettier because he accomplished so much in other spheres. However, Harkins' defensive aloofness allowed the church's enemies to fling the charge of hypocrisy at a religious institution always crying foul against so many criticisms while refusing to take advantage of honest chances to break bread with neutral or potentially friendly forces.[10]

Banigan, on the other hand, sought the wealth to finance his philanthropy in a primitive but far-flung global marketplace. As a scientist and hands-on inventor, he may have been accepted in a select, but tolerant society of "pluck and luck" practitioners of creativity who valued knowledge and imagination more than ethnic background. The constant travel and business negotiations within the cultures of foreign societies provided him with a broader perspective. Harkins, weighed down with encyclicals, plenary constraints, and church doctrine, marched over a narrower, pedestrian terrain. During a three-decade career, he immersed himself in social justice and social services for his variegated flock, creating a remarkable twenty humanitarian agencies in an urban terrain almost barren of such facilities. Historian Patrick T. Conley describes Harkins as having "made a greater and more enduring impact upon Catholicism in Rhode Island than anyone before or since,"[11] especially the establishment of Providence College in 1919. Although many of Harkins' institutional accomplishments came after Banigan's death in 1898, several important developments solidified their close personal relationship and set the stage for later projects and innovations. Still there is no escaping the fact that the rubber king played the quintessential free agent with few restraints in his business dealings. Bishop Harkins, weighed down by centuries of cumulative church law and lore, carried his own cross through the mean streets of Providence rather than the divine Providence founded by an ecumenical Roger Williams generations earlier.[12]

The Home for the Aged

His home parish memorialized Banigan after his death in 1898 with the words: "The full range of his extensive private charity will never be known."[13] The generosity must have been muted or camouflaged during his early career. Although the Bishop performed his wedding ceremony in 1860, that might have been due to John Holt's benefactions and influence. In 1881 the Little Sisters of the Poor, a French order dedicated to caring for the elderly, came to Providence. The aged among the Irish-Catholic population faced problems that combined the frailty and poverty common to most in that era. Originally the Sisters cared for a small group of patients in cramped quarters and actually begged for food and raiment. Banigan took an interest in the work of the Order and transformed that concern into his first major philanthropy. He purchased a piece of land in the Woodlawn section of Pawtucket in 1882, just over the city line from Providence. He bankrolled a four-story, state-of-the-art brick facility to accommodate 350 residents. At the dedication in May 1884, Bishop Thomas Hendricken, assisted by Vicar General McCabe, celebrated mass. The *Providence Visitor* beamed: "No better evidence of the growth of Catholicity in our state and city can be obtained than by the prosperity and increase of its charitable institutions." The newspaper also gushed in the unending parochial attempt to gain recognition and respect from the Yankee ascendancy: "The extent of the charities of the Church cannot fail to impress the Protestants as well as the Catholics."[14] The structure lasted almost a century until a new facility replaced the old quarters in 1979.[15]

Joseph Banigan's altruism and kindness pegged him as the Catholic Church's most influential financial backer in the state. He purportedly donated $160,000, about half the cost of the structure. Not surprisingly, the wherewithal for such kindness came after Banigan secured full control of the Woonsocket Rubber Company in 1882. He also employed his penchant for detail in charitable ventures just as in business. The four-story structure for the Home stood so tall that infirm nuns could watch the funerals of elderly residents across the railroad tracks at St. Francis Cemetery a half-mile away, where Banigan would eventually build the mortuary chapel for his family. Perhaps even then he knew that one day they would watch over him.[16]

Banigan did something else at the time that marked him as a person of exceptional character. He insisted that the admission policy to the facility not discriminate on account of creed, nationality, or gender. In fact an earlier venture apparently soured when the incorporators designed bylaws that included religious and racial bias. Residents under Banigan's rules had to be sixty years old, destitute, and possess good moral character. However, the sisters would not admit the demented. Although it is difficult to determine how far that open procedure actually went, the very concept was unusual and more in keeping with the spirit of Roger Williams than usually exhibited even by the Yankee descendants of the state's founder. Father Robert J. Sullivan, the rector of St. Mary's Church, made a very direct "thank you" to Joseph Banigan but couched his praise in a business metaphor.

> I know nothing about financial enterprises, but I venture to say that the erection of this Home is the best investment you could make of your money, the one that will pay the highest rate of interest, because it depends not on the fluctuations of the market and the tricks of speculators. You have made, as it were, God and chosen Ones your debtors and they will not forfeit their obligations. May the noble example set by you to-day find in every country generous imitators.[17]

Editorially, the *Providence Visitor* gushed that: "No event of this kind ever occurred in Rhode Island before." The diocesan piece concluded with an unusual embrace of Banigan's toleration: "Large benefactions have been made, but none that were free from the limitations either of creed or nationality—none so broad as to include within its benevolent design the Christian spirit upon which this new foundation rests."[18] Bishop Hendricken was already sick and absent from the dedication. The sentiment of openness expressed by the diocesan newspaper reflected his more ecumenical administration than the forthcoming ascension of the less liberal but brilliant tenure of Bishop Harkins.[19]

A Knight of St. Gregory

Banigan's largesse, with much more to come, earned him an astonishing honor in the bittersweet summer of 1885: Pope Leo XIII appointed him a Knight of St. Gregory, an international philanthropic order that theoretically made up the pontiff's honor guard. Pope Gregory XVI established the Order in 1831 for "gentlemen of proved loyalty to the Holy See who, by reason of their nobility of birth and the renown of their deeds or the degree of their munificence, are deemed worthy to be honored by a public expression of esteem." Knighthood in the Order remains, to this day, the highest distinction bestowed on the laity in the Catholic Church. Banigan became only the second American so recognized. The letter from Rome arrived during the strike wave at the Woonsocket Rubber Company. Surprisingly, the honor garnered almost no publicity in the local press although the Knights of Labor used the title to mock Banigan's perceived hypocrisy. "Giuseppi" Banigan's reputation crossed the Atlantic Ocean almost forty years after his departure from Ireland in 1847, and the recognition had nothing to do with "nobility of birth." The only real obligation that any honoree accepted was continued exemplary service to the Church. Banigan required no such reminder.[20]

The Catholic Orphan Asylum

The creation of a modern orphanage in Rhode Island marked another great contribution to social justice and Catholic charity. Banigan's interest in an asylum for youth reflected, once again, his concern for children. That anxiety and solicitude, shared by many others in the Irish-Catholic community, ran deep among the Famine generation. The "Great Hunger," with its attendant diseases, had decimated the Irish population, especially the young and very old—the usual victims in these calamities. Furthermore, general medical problems in childbirth and childhood, irrespective of any spectacular contagious disease, took a toll among individual families. Brothers and sisters, sons and daughters regularly perished as part of everyday life in most parts of the globe. Those emigrating from foreign shores faced a number of humanitarian crises that paralleled Old World problems. In the United States, and Rhode Island was no exception, Irish parents might leave a baby in a public space hopeful that a person of means would adopt the "foundling" (the popular term for deserted children in the nineteenth century). Many Irish families simply could not afford another mouth to feed and took drastic actions.[21]

The scandalous nature of child abandonment and the attribution of such infamy to Irish Catholics created great concern among the community's leadership. In fact, Bishop Harkins addressed this issue even before he tried to manage so many of the other social problems among his parishioners. The Church established the Rhode Island Catholic Orphan Asylum in 1851 in the Irish Fifth Ward as the population exploded with the arrival of Famine refugees. Eventually, the Diocese erected more substantial quarters in the heart of the Irish community in South Providence (part of Cranston at the time) in 1862, just outside the boundaries of the Fifth Ward. In the fall of 1887 Bishop Harkins, with Joseph Banigan in tow, inspected the then deteriorating and overcrowded facilities. Even the best efforts of the Sisters of Mercy, who toiled endlessly to keep the place running properly, proved insufficient. A lay organization, the St. Vincent de Paul Society, had longstanding activist ties to both the Order and the orphanage. Bishop Harkins and Joseph Banigan always welcomed greater rank-and-file involvement, especially at a time when the clergy was brutally overextended and the depth of Banigan's pockets neared its limit. Within two years of the inspection, two additions to the original structure doubled the size and capabilities of the asylum.[22]

Although the participants realized that the facilities served the Irish-Catholic community almost exclusively, the open enrollment policy still reflected the liberal nature of other such organizations that Bishop Harkins and Banigan established. The introduction to the new charter for the asylum, however, taunted local politicians: "It is impossible to estimate how many thousands of dollars have been saved to the cities and towns of the State, by the asylum, during all these years." The writer then lowered the boom: "No one will doubt the duty of the State in the matter. But we must go on as in the past, and depend upon the charity of a generous public."[23] The facility had accommodated an estimated four thousand residents since its inception at a miniscule cost of about fifty dollars a year for each child, thanks mostly to the voluntary sacrifices of the nuns.[24]

Banigan, as usual, became the point man for the money. At a dinner meeting with the bishop in Newport in September 1887, he pledged to donate $1,000 a year for a decade and organize an association of 100 supporters, each of whom would contract for $200 annually. Later he offered to increase his contribution to $2,000. Banigan's continued personal largesse also reflected the trouble he experienced lining up the limited available number of Irish-Catholic benefactors. James Hanley, the influential owner of the state's largest brewery, wrote that, "while I fully sympathize with the object of the movement, and will do what I can to forward the project, I do not feel that I ought to accept the position of Trustee, without first knowing more of what my responsibility would be."[25] Banigan put a positive spin on Hanley's demurer in a note to the Bishop, but he also admitted that his own older brother, Patrick T. Banigan, had backed out of the commitment altogether because "he is not sufficiently acquainted with finances to intelligently handle such an amount of money." Banigan displayed a bit of humor when he advised Harkins that "I presume to say that I was the wrong party to see him as his first question was 'Why don't you do it yourself?'"[26] The rubber king had probably gone to the well too often with the limited number of other affluent Catholic activists and frequently made up any shortages from his own pocket. Like most of the Church's endeavors, the construction consisted of a one-time set price, while the operational costs fluctuated. Banigan never hesitated in his own guaranty, frequently meeting the Bishop at the asylum to supervise or inspect building additions, something he knew about from his own business experience. A sizeable amount of the shelter's endowment was invested in the Woonsocket Rubber Company.[27]

One uncertainty in the operation of the orphanage, and other such facilities around the state, focused on the various ages of the occupants. Children under four years old presented special problems not easily addressed in an institutional setting for adolescents. The St. Vincent de Paul Society initiated a plan to establish an infant asylum separate from the older refuge. The Society, with decentralized chapters around the state, possessed the manpower, organizational skill, and middle-class funding necessary for such an important but costly venture. The new facility for infants opened in February 1892, and besides the usual disclaimer about ethnicity and religion, children would be admitted without regard to race as well. However, the asylum would *not* accept foundlings that had once plagued the Irish-Catholic community. The orphanage now sent abandoned toddlers to state authorities, probably a reflection of greater prosperity and awareness among Irish Americans to eliminate the practice. The building in the Smith Hill section of Providence quickly filled. In 1895 Banigan purchased a seven-acre estate on Regent Avenue for $22,000 and donated it to the Society for immediate expansion. He must have made a sigh of relief that a small army of volunteers filled the breech this time around. Banigan's four children donated the funds for the kindergarten in memory of their mother, Margaret Holt Banigan. Bishop Harkins certainly never rested in his quest to provide his ever expanding flock with the requisite charity and services they needed. He commandeered Joseph Banigan's free time from his duties at the rubber company whenever possible. The rubber king sat on the corporation board. The cornerstone ceremony, attended by an amazing four thousand supporters, included Edwin D. McGuiness, elected the city's first Irish-Catholic mayor in 1896.[28] Bishop Harkins lauded Banigan *in absentia:*

> It seems almost providential that this site was acquired for the St. Vincent de Paul Asylum. This was the first site selected years ago. At that time it was impossible to secure it, and now that we have it, I look upon it as a mark of especial favor of the hand of Divine Providence. This location was made possible by one man's magnificent generosity. He is not here to-day. We certainly regret his absence and would desire a place of honor on the platform for him. We all pray that he may be long spared to us, here in Providence, in the business world in which he takes such an important place, and especially in the world of religion, education and charity, in which he has given such a magnificent example of the proper use of wealth by his most generous and universal charities.[29]

St. Maria's Home for Working Girls

Bishop Harkins juggled dozens of ongoing projects, many of them high-end enterprises. Banigan had his fingerprints on many of them although his business ventures interfered with too many time-consuming charitable pursuits. When the prelate decided that Rhode Island needed a Catholic hospital, he set out to accomplish that goal. After only a few years of planning and building, St. Joseph's Hospital, staffed by the Sisters of the Third Order of St. Francis of Allegheny, New York, accepted patients in April 1892. The facility, located on Broad Street in Providence just outside the old Irish Fifth Ward, quickly filled. The hospital added another wing in 1895, constructed by the Gilbane Construction Company, one of the few Irish concerns eventually to parallel Banigan's wider success. He made significant contributions to the Catholic facility but also gave regular donations to the secular Woonsocket Hospital and the Providence Lying-In Hospital for childbirth, another reflection of his concern for the very young regardless of religious affiliation.[30]

Banigan created another social service agency in 1895, with the help of his second wife, Maria Conway Banigan, a New Yorker, whom he married on 4 November 1873. They worked in close cooperation with the Diocese of Providence to establish St. Maria's Home for Working Girls. Gilded Age morality, regardless of the religion involved, winced at the condition of domestics, who often lived in the very homes they serviced. Society, at the time, feared for the welfare of these young women, often immigrants, who might suffer sexual exploitation at the hands of an unscrupulous employer. Another anxiety, especially within the Irish-Catholic community, focused on host families that might induce a servant to convert to their religion.[31]

In Providence, the Women's Christian Association (predecessor to the YWCA) offered rooms as early as 1867 to those pursuing a career. This predominantly Protestant establishment proved inhospitable to young Catholic females who sought temporary quarters there, although the number of Irish women in such a professional situation would have been small at the time. Apparently, the YWCA authorities told Catholics to have the Church build a facility of their own. Bishop Thomas Hendricken drew up preliminary plans to do just that before his death in 1886. Mrs. Banigan took the idea and drew in other supporters at meetings in her home. The founders envisioned a self-sustaining facility that charged a minimal fee for room and board for a mixture of Catholic women: domestic servants as well as single, female factory workers, nurses, teachers, and others. They also made provisions to care for unemployed, sick, injured, and even retired women. Residents would live in a family style setting and have a say in the running of the place. Although the home ostensibly operated for "indigent working women without distinction of creed or nationality," similar to the policy in effect at Banigan's other facility, the Home for the Aged, the tenants covered a wider class spectrum.[32] The pronunciation of the facility initially was the Gaelic term for Mary, Mariah. Over time the name changed to the more common Maria and eventually Marie.[33]

Mrs. Banigan's committee (including her stepdaughter, Alice) comprised affluent Irish-American women whose status derived from their husbands' standing and who served as their conscience in the wider community. They secured an agreement from Joseph Banigan to provide the construction costs if the organizers would raise $15,000 through a series of fundraisers for the maintenance and support of free apartments for the disabled. Twelve women, from wealthy families, also contributed $1,000 each to secure a retirement room in the proposed facility on fashionable Governor Street on Providence's East Side, not far from the Banigan residence. The arrangement almost paralleled modern-day provisions for a condominium. Banigan's wife and Bishop Harkins worked on an agreement to have the facility staffed by the Sisters of St. Francis. The prelate remarked in his diary in November 1894 that "Mrs. Banigan is collecting for the Home."[34] The St. Maria Society, the mostly lay committee to oversee and manage the place, consisted of twenty-three women and Joseph Banigan, serving as auditor. His wife engaged a prestigious architectural firm to design the three-story, French Renaissance building, which accommodated one hundred residents and the staff of the Sisters when it opened in January 1895, complete with a small chapel. Banigan paid the $50,000 for construction and provided a stock trust from the $15,000 the committee solicited, as he usually did with institutions he developed. Interestingly, he invested the money in shares of the American Wringer Company, the old Bailey establishment in Woonsocket that Banigan had transformed into a monopoly. At the time, the rubber king also provided funds to various boarders who were homesick and wanted to return to Ireland. He even paid passage for relatives who wanted to join them here. He left another $25,000 for the Home in his will.[35]

The thirty-three single women who resided at St. Maria's that first year, mostly between the ages of fifteen and twenty-four, earned up to $27.00 a month in outside wages. In April 1895 the Home collected just over $400.00 from the boarders. By 1902 the number of residents climbed to eighty-four. Some actually retired there after a lifetime of toil and eventually transformed the Home into a facility for the elderly. The Sisters of St. Francis left in 1973 and the building lay vacant for almost twenty years until a $5,000,000 renovation, completed in 1996, offered fifty-three units for low-income elderly residents. The local Preservation Society called the changeover miraculous. Banigan

A postcard view of St. Maria's Home for Working Girls, a Banigan venture (Author's collection)

adroitly insured the survival of his charitable institutions with endowments that assured their longevity.[36]

Catholic University

The dedication of the Home on 15 January 1895 featured almost as much patriotic as religious fanfare. One hundred and fifty attended. Fr. Thomas Grace of Olneyville, an area that was the vortex of the city's industrial might and union turmoil, preached a sermon emphasizing labor peace and Catholic charity by "the blessedness of devoting wealth, rightly acquired, to further the progress of the world and serving to make it better by devoting it to work of practical charity such as the St. Maria Home."[37] Banigan turned over the deed, keys, and an insurance policy. He also presented a hefty check to a special guest, Bishop John J. Keane, president of Catholic University in Washington, D.C. Keane was a close friend and confidant of Bishop Harkins in Providence who enthusiastically endorsed the establishment of the university as a place for graduate religious study. Bishop Keane visited Providence on several earlier trips. In July 1888 he had made a personal plea and co-signed a circular letter with Bishop Harkins for funding for the school, which opened a year later. Harkins gave generously himself. He served on the university's management and library committees and, in 1903, joined the Board of Trustees. Keane revisited the city in July 1889 to speak at the consecration of the Cathedral of Sts. Peter and Paul. The close relationship between the two Bishops brought Banigan into the Catholic educational orbit with some words of encouragement from James Cardinal Gibbons, a pro-labor, Irish-American prelate. Banigan served in many ways as the financial sponsor for Bishop Harkins' projects and, in this instance, followed him to an out-of-state venture.[38]

Bishop Keane's annual report in 1895 noted simply that "We have received from Mr. Joseph Banigan, of Providence, R.I., the sum of $50,000 for the endowment of the Joseph Banigan Chair of Political Economy."[39] The following year Catholic University appointed Banigan to the school's board of trustees. Banigan promptly announced, at the dedication of a new hall on the campus in the nation's capital, that he would make an annual donation to the library. He pledged $1,000 and augmented that with a stock investment from the Werner

Publishing Company that yielded $4,000 a year for the purchase of books. He agreed to continue that arrangement until the sum reached $50,000. "Mr. Banigan," the Catholic University *Bulletin* lionized, "has the gratitude not only of the Board of Trustees, but of every professor and student in the Institution."[40] The university listed $8,000 from the "Banigan Library Fund" by the end of 1897 and purchased 250 law books as well as volumes for "good scientific work." Banigan apparently remained an inventor at heart. The university's library committee must have rejoiced at Bishop Harkins' involvement that, more than likely, drew in the rubber king.[41]

Banigan, however, added one surprising proviso to the book fund: "Should I die before the dividends turned over to you equal the amount of the investment, this dies with me and will not be binding on my heirs."[42] Banigan unexpectedly succumbed two years later and the donation short-circuited. After his death at the age of fifty-nine, Banigan's wife, Maria Conway, contested the will because too much went to charity. The trustees of the estate made an out-of-court settlement that, no doubt, favored her interests. The tension between husband and wife over Banigan's charitable largesse probably had begun to percolate before the insertion of the unusual language (at least for Joseph Banigan) concerning the library funds. Even though Mrs. Banigan played a key role in the funding and construction of St. Maria's Home earlier, she may have felt that her husband went overboard with his contributions when the charity crossed the state boundary and threatened to grow out of control. The language to return the Werner Publishing Company stock back to the estate may have been a temporary compromise between spouses at the time that boiled over once again after the reading of his will.[43]

Brown University

Banigan, with only a year of elementary school to his credit, certainly believed in the opportunity to learn despite his own vast achievements without a formal education. While his interest in and kindness to Catholic University made great sense given his own background, a smaller but more controversial commitment—literally in his own backyard—raised eyebrows. Brown University, perched on the east side of Providence overlooking the downtown area, had its roots in the colonial era. Like many such institutions from that period, the school trained ministers, in this case Baptists. At times anti-Catholic rhetoric crackled from the campus. However, by the Gilded Age the forces of change and liberalization made their way up College Hill. President E. Benjamin Andrews, a minister himself, while not ending the hostility between Yankee Baptists and Irish Catholics, dampened the fire, even though antagonism reappeared on several other occasions later in the early twentieth century. In 1893 Bishop Harkins even sent a letter of recommendation to Andrews for a priest to attend the school. Andrews paid a price for his toleration, as one alumna took the university out of her will because the Brown president recognized Catholics as Christians![44]

In 1896, a year after he presented Bishop Keane with the famous $50,000 endowment, Banigan informed the governing body of Brown University that he would donate five $1,000 bonds, drawing 6 percent interest, for needy students to attend the school. Although the contract didn't mandate that the scholarships should go to those with an Irish or Catholic pedigree, he insisted on authorizing the grants to selected students. Before his donation, Brown admitted only a handful of Hibernians. "In the last decade of the 1880s," according to a retrospective article, "newly arrived students of Irish heritage were not invited, presumably because of their ethnic background and religion, to membership in Brown's collegiate fraternal Societies."[45] The thirteen matriculating Irish-Catholic students (ten from Rhode Island) gathered at Room 3 at Hope College on the campus to form their own fraternity in 1889. Originally called Phi Kappa Sigma, or the Fraternity of Catholic Students, the group did not seek a state charter until 1902, although the society spread to other colleges and universities before that. Room 3 became the informal meeting place for these undergraduates, who might drop in "for friendly conversation, perhaps even in Gaelic." The Irish language was probably learned at local classes, as most of the students seemed to be at least second- or third-generation Americans.[46]

The Irish-Catholic brotherhood enjoyed an informal network of support from the few earlier graduates of similar training and experience. Providence businessman Joseph M. Harson, owner of a downtown millinery shop and an East Side resident, opened his home to the organization and other sympathetic alumni. A fraternity article described Harson as "a man intensely loyal to Brown and devoted to the welfare of the Catholic students."[47] He served on the Board of Directors of the Rhode Island Catholic Orphan Asylum with Banigan and may have been the conduit for the latter's scholarship support. Frequent visitors to gatherings at Harson's home included Edwin McGuiness, elected the state's first Irish-Catholic general officer (secretary of state) in 1887 and mayor of Providence in 1896, and George J. West, a city politician and ethnic activist. John J. Fitzgerald, one of the fraternity's pioneers, became the first populist mayor of Pawtucket in 1900. In 1902 he actively supported a blockbusting statewide strike by motormen and conductors, and later became a major defense lawyer. Thomas P. Corcoran, another of the thirteen original fraternity members and also from Pawtucket, influenced his son's involvement in politics. Thomas G. "Tommy the Cork" Corcoran, advised President Franklin Roosevelt during the New Deal. Joseph Kirwin, also a charter member of the fraternity, may have been the first Brown graduate to become a Catholic priest.[48]

Although not one of the founding fathers, James H. Higgins joined Phi Kappa Sigma later, at age thirty. Higgins became the state's first Irish-Catholic chief executive in 1907. The "boy governor" had enrolled at Brown in 1894 and, although poor, "the earnest young Celt" impressed Brown President Benjamin Andrews, a Protestant minister, who promised to subsidize his education. Higgins may have inspired Joseph Banigan to provide those supposedly undifferentiated scholarships in 1896. The future governor won the university's most prestigious honor at the time, the Hicks prize for debating. Both he and Fitzgerald attended Georgetown Law School. Higgins succeeded his classmate as mayor of Pawtucket before becoming governor. After short political careers they formed one of the most potent law partnerships in Rhode Island. A fraternity brother who became a newspaper reporter wrote in the *Boston Sunday Globe* that Higgins was "The Plumed Knight, The Young Lochinvar, The Boy Demosthenes and everything that they apply to earnest crusaders." Ten of the thirteen founders of the Catholic fraternity became lawyers or physicians.[49]

The Mormons

Banigan's gift seemed a bit out of line, with so few Irish Catholics to support at the university, but he took the high road, preferring to see the cream of Irish-American society in Rhode Island rise to the top rather than curdle in second-class institutions. Banigan added a wrinkle to his unpredictable experiments in diversity. The $5,000 note he assigned to the university was secured by the president and vice president of the Mormon Church in Utah and by that state's influential senator, Frank Cannon. Banigan, near the end of his life, participated heavily in projects with the Mormon Church, including the sugar beet industry and the development of electrical energy. In fact, the five bonds assigned to Brown University carried the name of the Pioneer Electric Power Company in Utah, an investment favorite of the rubber king from Rhode Island. Banigan's financial dealings seemed to recognize profit only, regardless of extraneous features. The imprimatur of the Mormon Church may actually have surprised the Baptists at Brown more than local Catholics, who were getting used to Banigan's idiosyncrasies. His unorthodox investments proved unimpeachable— at least during his lifetime—because of his stellar record with his own kind irrespective of the strike of 1885. His religious toleration resurrected the ghost of Roger Williams despite the supposed narrowness of life within Rhode Island's Catholic envelope.[50]

In 1889 Banigan rented a private set of railroad cars to take his family on a tour of the Grand Canyon, one of those Gilded Age excursions that industrial magnates seemed to favor. On the journey Banigan may have visited Utah, where the Mormons, operated a primitive footwear production center of their own and also sold thousands of pairs of Banigan's quality products. *The Boot and Shoe Recorder* remarked in a story that "They handle a large line of Woonsocket rubber goods."[51] Zion's Mercantile Co-operative Institute, situated in Salt Lake City, had three branch outlets in Provo, Ogden, and Logan, which sold $5 million in department store articles annually. Banigan, a peripatetic tourist who enjoyed mixing business with pleasure trips, may have found the Mormons a welcome relief from the dog-eat-dog world of business and eastern society.

He may also have learned that at the time of the Great Famine in Ireland, the Mormons attacked the wealth of the Anglican Church and brilliantly organized an exodus from Liverpool to the United States and Utah at the same time the Irish fled their homeland from the same English port. Utah senator Frank J. Cannon, son of one of the ruling disciples of the Mormon Church—George Q. Cannon, who left Liverpool at age thirteen in 1842—paid a visit to Banigan at the southern Rhode Island resort of Narragansett Pier in 1896. Cannon, probably building on Banigan's favorable impression of Mormon enterprise, initially sought a controversial investment in the sugar beet industry that failed to unify even the leaders of the Latter Day Saints. Behind the scenes, they argued about the advisability of the Church's economic activism and the wisdom of seeking capital from a gentile, a highly unusual move on their part.[52]

Banigan invested $360,000 in the Utah Sugar Company, which also provided irrigation in that parched western area. Joseph F. Smith noted that:

> we met a gentleman from the East, who came out here to Utah on a visit and went down and took a look at the sugar factory, and at the fields of growing beets; he observed the industrious character of the people, and he said to himself, and to these men here, "I have sufficient confidence in your integrity and in your intelligence and wisdom, to invest at least one hundred thousand dollars in the purchase of your bonds."[53]

George Q. Cannon, the senator's father, admitted to convincing Banigan that "in the mountains there was a conservative element that could be relied upon in the days of trouble. The [Mormons] would not organize into mobs, they would not raise riots, they would not be carried away by the ridiculous ideas which find circulation from time to time throughout the country."[54] Banigan may have yearned for such respite even a decade after the 1885 strikes at Millville and the brutal competition within the rubber industry.

He made further investments in Mormon enterprises, purchasing $1,500,000 in bonds for the Pioneer Electric Power Company in Ogden, three-fourths of the entire capitalization of that enterprise, which eventually developed street railway services. Banigan said, "You people who will not actually have invested a dollar of money will have $1,700,000 of stock in a paying enterprise. It is only my absolute confidence in the . . . honest character of your people which prompt [*sic*] me to make an offer."[55] His interest loosened further capital from English sources, insuring the success of the venture. Upon his death in 1898, the Mormon Church lionized "the eastern financier" who allegedly had a portrait of Brigham Young, president of the Mormon

Church, on his office wall. The *Deseret News* not only carried a sensitive death notice of Banigan but uncharacteristically reproduced obituaries from the *Colorado Catholic* and the diocesan *Providence Visitor.* Banigan, who apparently could have collected an extra $200,000 from his investments there due to a technicality on capital exemptions from outside the state, refused the bonus and became something of a hero in Mormon society at the time. When the Mormons opened a Sunday school in Providence in September 1898, just weeks after Banigan's death, local papers and churches attacked them unmercifully. Although there is no evidence that Banigan endorsed, encouraged, or embraced that religion, he certainly would have discouraged such hostility, given his tolerant business mindset. Brigham Young's portrait seemed to be nothing more than a show of respect to important clients who happened to be part of a religious sect that seldom garnered any respect or tolerance from the outside world in that era. Banigan may also have felt a kinship because of the exodus and exile of Mormons and Irish from England in the 1840s, albeit under very different circumstances. Furthermore, Banigan decorated his offices with so many artifacts, curiosities, and knickknacks that it resembled a nineteenth-century cabinet display at an historical society rather than a traditional business setting. The Catholic Church issued no known public or private criticism of Banigan for his relationship with the Mormons, although privately there may have been some hard feelings.[56]

If Banigan's picture of Brigham Young in his office failed to aggravate Catholics, his friendly and hefty symbolic contributions to the establishment of the First Presbyterian Church in Woonsocket and of a Jewish synagogue, Temple Beth-El, in Providence probably caused no seismic eruptions either. The rubber king had turned into a nineteenth-century ecumenist.[57]

Shanty and Lace Curtain Irish

Banigan's encounters with death and disaster did not differ appreciably from those of his countrymen. They too lost loved ones, suffered injuries and death in industrial accidents, behaved in ignorant and primitive ways, and cried uncontrollably at special masses for the departed. Some, like the rubber king, lived luck-blessed lives in the shade of their shamrocks. To others, the stench of the Famine and cheap liquor clung interminably. Whatever the circumstances, the prosperity and fame of a Banigan, Gilbane, or Hanley in business or a McGuiness, Higgins, or Fitzgerald in politics, provided hope against all odds that the Irish immigrant group and their progeny could eventually succeed, anchored by the rock-solid Catholic Church. The benefactions of the elite or their organizations not only offered employment but, more important, the security that secular salvation was just around the corner just as spiritual redemption was already promised by the Church.

The advances of the lace curtain Irish did not happen in a vacuum. Banigan's Brown University bequest, backed by investments in Utah, did not happen independently, outside the temper of the times. For example, in an amazing turn of events, the managing editor of the *Providence Journal*, Alfred M. Williams, publicized the Gilded Age Irish literary renaissance during his stewardship of the paper from 1884 to 1891. Williams was a farm boy from Taunton, Massachusetts, who attended Brown University himself for a couple of terms before joining the Civil War in the volunteer infantry. He wrote articles about the conflict for the crusty publisher Horace Greeley and his *New York Herald,* and went to Ireland in 1865 for several months to describe the Fenian revolts. In a circus of errors, the British authorities arrested him for a week upon his arrival, which only endeared him to readers back home. While in Ireland, he compiled information for pithy articles and began to collect literary and political memorabilia. Williams became a partisan of things Irish. He joined the *Providence Journal* as a reporter in 1875 and continued to acquire Irish cultural tracts, folklore, and poetry while publishing the works of avant-garde Irish writers in the newspaper's Sunday edition, which he instituted several years later. The turnabout from his predecessor, arch-nativist Henry Bowen Anthony, must have been as shocking to Irish Catholics as to the Yankees, because his embrace of Hibernian life stretched all the way to support for Irish Home Rule. Ironically, he succeeded Anthony, who ran the *Journal* and the state for almost half a century, reaching back to the days before the Dorr Rebellion. A second trip to Ireland reinforced Williams' earlier proclivities and cemented his relationship with up and coming writers like William Butler Yeats, who wrote that the *Providence Journal* became his "only regular and certain paymaster."[58] Upon his death in 1896, Williams bequeathed his impressive primary collection of 2,000 books (including his own study of Irish poetry), manuscripts, and broadsides to the Providence Public Library, one of the greatest accumulations of original holdings outside of Ireland.[59]

The flowering of Irish-American society decorated the Gilded Age in Rhode Island. Henry Bowen Anthony's replacement by Albert Williams demonstrated the dramatic impact such a turn-around could have in the influential pages of the state's leading newspaper: from verbal caricature to cutting edge Irish literature. The cracking of Rhode Island's most prestigious institution of higher learning by Irish-American students, financed by Banigan scholarships, underlined the emerging ethnic elite that could battle the best of the Yankees in scholarly pursuits at Brown University. Similarly, the increasing number of Irish-American politicians, who achieved virtually every office the state had to offer, especially in the early twentieth century, reflected a massive change in the electoral landscape that had more to do with the enfranchisement of native-born Irish Americans in Rhode Island almost a half century after the Famine than with any change of heart by Yankee Republicans, or Democrats for that matter. Voters passed the Bourn Amendment to the state constitution in 1888, introduced by state senator (and former governor) Augustus O. Bourn of the National Rubber Company in Bristol, which eliminated the albatross of property qualification to vote in high profile elections for mayor or governor. Naturalized citizens, at least on the face of it, possessed the same right to cast a ballot as the native-born. Cagily, however, the Republican Party shifted the property threshold requirement and applied it to all, native and foreign born, to aldermanic elections and financial referendums where the real power rested. The GOP maintained a stranglehold on restrictive apportionment in the state senate that preserved conservative, rural power for almost another fifty years. Historian Patrick T. Conley estimated that the constitutional change coincided with a demographic shift that catapulted the local Irish—native and foreign-born—over the 100,000 mark in a population of just over 300,000. The Republican Party, with an eye on the vanguard of the new immigrants who might balance the growing power of Irish Americans, cynically championed the electoral rights of these newcomers after fighting against the same liberalization for the Irish for over six decades. Still, as Conley wrote, the influential Hibernian politicians were "merely the highest blades of a lush green lawn of local politicos that have sprung from the 'auld sod.'" Patrick J. McCarthy, a Famine refugee from Ireland, came to the United States about the same time as the Banigans. He settled in Rhode Island and served as Providence's only immigrant mayor ever, in 1907 and 1908. Starting in 1913, Irish Americans would serve a sixty-year, unbroken tenure in that office, almost like the local bishops in the Catholic Church.[60]

At first blush the veterans of the Famine, with their children and grandchildren, seemed to move ahead in unison, breaking free from the dungeon colony of Providence's Fifth Ward and other similar enclaves around the state. They scaled the walls of academia and eventually built their own institutions of higher learning just as they had established a thriving, lower-level parochial system. For a time, at least, the news and editorials of Anthony's diabolical *Providence Journal* appeared to have been dampened by holy water. Irish-American politicians walked the halls of power throughout the state and, in some places, received Yankee supplicants requesting favors in an amazing role reversal. The rank and file Irish took to politics, especially within the Democratic Party, as strongly as they frequented church. The ragtag army that escaped the pestilence in Ireland and overcame the bigotry here seemed ready to seize the day. Charles Carroll, who wrote a multivolume history of Rhode Island that was the first to include the contributions of immigrants in general, claimed "The politics of America intrigued the Irishman; he longed to participate." The cultural, educational, and political advancements seemed the proverbial tip of the iceberg. So many years of disdain toward Irish Catholics seemingly entered the genetic code of the state's rulers and their followers. Unfortunately, the meltdown was only a sign of an Indian summer. The local Irish-Catholic community required more time to overcome the institutional prejudice and, although neither the leadership nor the rank and file wanted to admit the truth, they sometimes perpetuated the bias by their own shortsighted actions, like that of the "drunken Irishwoman" who was ejected from a Pawtucket horsecar for frightening other passengers. A letter from a student to his uncle in the same city mentioned that "I wish you were here to watch this gang of Irish go to church all dressed up in green and they put on airs enough," adding for emphasis, "You bet they do." Even as late as 1888 the Irish accounted for approximately 7,400 of the 8,000 arrests for public intoxication in Rhode Island, although percentages for other offenses varied little from those of other ethnic groups. A Providence diarist wrote in February 1889: "Went down to Park's Continental Band Fair in the morning, regular, rough Irish crowd, all dancing and smoking cheap cigars."[61]

Alfred Williams, the patron of the Irish literary renaissance locally, served as editor of the *Providence Journal* for several years before retiring in 1891. After his departure, Hibernian writers seldom graced the pages of the newspaper. The only renaissance after his exit was a recrudescence of headlines and news that always seemed to denigrate Irish Americans in some editorial form: the paper insultingly raising an orange flag on top of its headquarters on St. Patrick's Day. The rising of Irish literature out of the caricature of the boglands merited great attention, as did Williams' publication of so many articles, essays, and poems despite the elite character of the material. The scholars at Brown, though their ringing success demonstrated that the cream of the crop could indeed make it in the Ivy League, served as models for those behind them. They also became targets. The *American Citizen*, a weekly anti-Catholic newspaper based in Boston, printed many notices from Rhode Island. One letter in February 1896 took Brown President Benjamin Andrews, "the Romanized professor-president," to task for several things including a visit by Bishop Keane, the rector of Catholic University and recipient of Banigan's $50,000 contribution. The writer complained: "what has Brown had from the Romans to be grateful for? Has one of that faith established a free scholarship or endowed a chair? Has there ever been a donation of any character, or for any purpose, in harmony with the establishment?" There was also a swipe at Edwin McGuiness, the pioneering Irish-Catholic politician who supposedly "Hiberniacized" electoral offices in Rhode Island.[62] The very syntax of the correspondence was similar to the Wilkinson letter questioning the existence of any Irish-Catholic contributions to industrial progress. In both cases, Banigan answered the call.

Still the Hibernian presence at Brown University provided an important symbol at the grassroots level. A comparison between Yankee and Irish-American students in Providence between the ages of thirteen and fifteen uncovered why. In 1880 more than three-fourths of Yankee children in that age bracket attended school and less than 15 percent worked full-time. Just over one-third of Irish immigrant students, on the other hand, enrolled in classes; almost one-half toiled in industry. By 1900 the figures for Irish immigrants and the children of the Irish in school showed a gain of 20 percent while the Yankee numbers dropped slightly.[63]

Upward mobility placed a great strain on the relationship between Irish-American leadership and the rank and file, although no one, understandably, wanted to discuss the matter. As Banigan empowered his rubber operatives with decent wages, working conditions, and a degree of independence, those benefits allowed the workers to chase their own destiny at Banigan's expense rather than that of some Yankee mill baron. The Woonsocket Rubber Company, because of its size and ethnic workforce, was atypical in Rhode Island, although the Knights of Labor grappled with many a Yankee kingpin in that period as well. In other areas, shopkeepers like Joseph Harson (who later moved to New York) assisted Irish-Catholic students at Brown University, but prudently remained obsequious to his Yankee patrons at the downtown millinery store. Those Irish, Ivy League students who became lawyers and physicians respected their ethnic clientele however far away they lived from the original ghetto. Irish politicians, whatever the size of their ego or political machine, had to stay close to the pulse of the neighborhoods and their denizens. And even the Church, as the bureaucracy increased and the forces of centralization took over, still allowed some leeway to the dynamic parishes and priests who were the local face of the institution.

Soon the pride and prejudice that accompanied a people's ascension would bifurcate the community as it had with Banigan and his Irish workforce in 1885, as detailed in the novel *Our Own Kind* about Hibernian life in Providence, Rhode Island author Edward McSorley.[64] The years after the walkout in Woonsocket and Millville proved to be the greatest in Banigan's career. He punctuated his business career with philanthropic endeavor and climbed the corporate ladder to heights seldom reached except by Irish-American ironworkers preparing the skeleton of some new-fangled skyscraper, like the one Banigan built in Providence near the end of his life in the 1890s. Banigan, and other Irish citizens, drew strength from their ordeals and moved ahead. And although they marched at a different pace, there would be no retreat.

RUBBER KING AND RUBBER WORKERS

· ·

*"Whether such a man [Banigan] shall be approbated and
encouraged when he not only attempts, like a highwayman,
to extort an unjust allowance, but seeks to do it by vilifying
from the start, — men who are as much above him morally
as the sun is above a lamp-post."*
*— Plaintiff's attorney in a disagreement over the price of
delivered crude rubber from a Yankee shipper to Woonsocket,
June 1887*

*"Seest thou a man diligent in his business, he shall stand
before kings."*
*— Rhode Island Governor "Honest" John Davis, toasting
Joseph Banigan at the opening of the Alice Mill in
Woonsocket, October 1890*

After the 1885 strikes by the Knights of Labor, Banigan seemed disinclined to worry about revenge in the short run. Like so many of his
corporate contemporaries, he took a long-term approach to downgrading the workforce by automating the workplace with modern machines that eventually replicated the dexterity and mastery of once irreplaceable artisans. The ongoing struggle between capital and labor
was also a part of the larger battle among footwear manufacturers,
who searched endlessly for a competitive edge along the fault line of
vertical monopoly, from the purchase of raw rubber to the wholesale
and retail sales of boots and shoes. As Banigan and his Irish employees resumed work at their customary places, the operatives seemed to
have insured that the weak link in the production chain would not be
the price of wages, as they tried to eliminate that factor from compe-

tition altogether. The owners, for the time being, would have to seek organizational savings elsewhere.

The rubber king entered the most productive period in his life, fully financed and highly confident, despite the bare-knuckled strike. He would claw his way to the top of the rubber industry, pay homage to the faith and ethnicity of Ireland, secure his regional customer base, and press forward to a larger landscape as a global merchant, a direction that he had already pioneered in the industry. At the same time, he initiated new investment ventures that allowed him to redouble his charitable endeavors. Bishop Harkins hailed the rubber king's philanthropy and publicly expressed the hope that Banigan enjoyed longevity in "the business world in which he takes such an important place." Some of the material benefits of his labor also found personal expression in a spectacular East Side estate that rivaled most Newport "cottages," the understated moniker for the fashionable palaces of the wealthy in that local resort town.[1]

By March 1886, the Woonsocket Rubber Company produced a record number of boots at the Millville plant, 650 single pairs in a day. A week later the company closed to observe St. Patrick's Day. At the end of the year the baking rooms operated all night long. The hyperactive pace prefaced the vast changes that transformed the industry in this era and led to the establishment of a rubber cartel several years later. The takeoff had less to do with the cutthroat competition in fabrication among the eastern manufacturers of footwear than with the commercialization of raw material in the Brazilian jungle along the Amazon River. The rising price of rubber, still gathered in a primitive fashion but now controlled by savvy locals and British investors, wreaked havoc with the industry in the United States. Joseph Banigan took action to dominate the supply side of the product and, by doing so, enhanced profits at the final stage of production. His actions provoked fierce opposition among other American rubber importers and helped trigger the birth of the rubber monopoly that ultimately controlled and equalized prices for all its members. Although Banigan initially stayed out of the cartel's fold, he prudently analyzed the infant trust's course of action. He eventually joined the fraternity after brokering a stunning price for his Woonsocket empire. As part of the deal he became the president of the United States Rubber Company and, *de facto,* king of rubber.[2]

A WEEKLY JOURNAL OF PRACTICAL INFORMATION, ART, SCIENCE, MECHANICS, CHEMISTRY, AND MANUFACTURES.

Vol. LXVII.—No. 24.
ESTABLISHED 1845.

NEW YORK, DECEMBER 10, 1892.

[$3.00 A YEAR.
WEEKLY.

MANUFACTURE OF RUBBER SHOES—INTERIOR VIEWS OF FACTORY OF BOSTON RUBBER SHOE CO., MALDEN, MASS.—[See p. 374.]

Production scenes from the Boston Rubber Shoe Company in 1892, very similar to operations at Banigan's plants. The graphics depict the ongoing changeover from artisan work to partial mechanization. (Scientific American, *12 December 1892*.)

The Roots of Monopoly

The rubber industry had early incentives to conglomerate almost from the dawn of footwear production, although time and circumstances continually changed the configuration and the rationale. The global search for a scientific formula to create a workable rubber product ended in 1839 with Charles Goodyear's discovery of vulcanization. That achievement, like the perfection of so many other products in that era and others, led some companies and in-house inventors (including Banigan early in his career) to attempt to sidestep any iron-clad patents with alternative processes. Any number of pairs of shoes and boots got worn out by a small army of lawyers walking across countless courtroom floors, including the U.S. Supreme Court, arguing for and against the legitimacy of Goodyear's patents and substitute methods in the 1840s. Six companies, the Goodyear Associates and Licensees, paid a fee to the inventor for the rights to employ his formula in the same period while fending off unauthorized infringements. The firms chipped in 3 cents for every pair of boots or shoes they produced to finance a common treasury to prosecute encroachments and even fine their own members for infractions. Furthermore, and perhaps most important, the companies set prices, discounts, and production quotas. They arranged to meet annually.[3]

By 1853 the group consisted of eight concerns that discussed the possibility of merging their businesses into one body: "We should then have in reality what we now have but in name, one interest and no competition."[4] They understood, albeit in a primitive way, the rewards of consolidation, standardization, the collective purchase of raw rubber, economy of scale, and the end to duplication of styles. The parties failed to reach a complete agreement on most of these complex issues. Difficult problems, involving processes or just the grating of personalities, impeded the ability to form a mature combination of diverse firms in that time period. However, the inability to reach a full-scale settlement did not quench the thirst for greater cooperation on a piecemeal and effective basis. The group and its successors, five of whom eventually lasted long enough to form the nucleus of the United States Rubber Company in 1892, continued to work cooperatively in a quasi-monopoly fashion. They distributed an annual newsletter to other firms (after the expiration of the Goodyear patent), pushing to fix prices or limit footwear production. This early precartel activity came long before the establishment of Banigan's Woonsocket Rubber Company in 1867. Yet Banigan, after the labor strife in 1885, loomed as a key member of subsequent cooperative activities, playing, at times, the role of both Jekyll and Hyde, depending on his own agenda.[5]

In 1882 another precursor toward monopoly control of the industry came from the next genius in the field. Dr. B. F. Goodrich issued a call to incorporate all rubber related firms, including many outside the flagship footwear producers, to control the acquisition and price of rubber in Brazil. Goodrich also pleaded that, at the time, the "business is unstable, liable to great fluctuations, and does not yield a reasonable return on the investment; all are constantly tempted to disregard cost of production while endeavoring to hold customers."[6] Goodrich, despite his prescient analysis, failed to bring the rubber businesses together. However, his concern over the price of crude rubber would indeed forge a degree of solidarity between the jealous competitors. Ironically, supply and demand played hide and seek with one another in the Gay Nineties. The popularity of the bicycle craze and its primitive tires jacked up the cost of raw material. The Rubber Boot and Shoe Manufacturers' Association formed in 1889 to tackle these expense problems at the front of the production process. As Goodrich was working toward consolidation, Banigan was finishing the world's most modern boot-making facility at Millville, certainly a disincentive for his participation in any rubber syndicate, as he stood astride an unmatched asset. Furthermore, the rubber king already planned to expand his empire. A year after the Goodrich appeal Banigan purchased a one-third interest in the Hayward Rubber Company in Colchester, Connecticut, and sent a trusted foreman to superintend the plant. He also owned stock in the Goodyear facility in Naugatuck, Connecticut.[7]

Earlier in 1883, Banigan had joined a national effort to protest the high cost of crude rubber (and eliminate overstock). The industry, including the Woonsocket and Millville facilities, temporarily suspended 1,000 operatives for four months. An estimated 10,000 workers, mostly in other New England rubber plants, suffered the same fate. Banigan used the down time to make repairs, but he progressively paid his workers half-wages during this time, to insure their immediate return when the factories reopened, in order to fend off other companies from stealing his skilled workforce. When workers resumed opera-

tions in May 1883, the Woonsocket Rubber Company recapitalized at
$1,200,000. If Banigan drilled his competitors he also could be a gra-
cious host, entertaining two top officials of the very independent
Boston Rubber Shoe Company with a tour and banquet. Elisha S.
Converse, one of his guests, reigned as the godfather of rubber production at the time and shared a philanthropic agenda with Banigan,
although of a more secular nature. The high price of crude rubber
still bedeviled the manufacturers. At a rare public presentation, Banigan spoke to the Woonsocket Business Men's Association in January
1887, informing his listeners that most crude rubber had once gone
directly to England before being reshipped to the United States, a latter-day mercantilism. Now the material came directly to America but
still in British ships. As he spoke, the price hit 83 cents a pound, a far
cry from the 51 cents a pound bonanza he secured during the 1885
strike. Despite any reservations he might have harbored, Banigan
joined the rubber combine in June 1887 in order to purchase rubber
directly and eliminate costly import surcharges, his usual procedure
for increasing profits and managing his own destiny, unbeholden to
middlemen.[8]

A Court Contest

Banigan had a longstanding grievance concerning an added cost to
the price of raw rubber. He often complained that the product arrived
caked in mud and other detritus from the long journey out of the
Amazon River in Para, Brazil. Heavily soiled rubber could increase the
cost of the material substantially, although industry producers probably chalked it up as the price of doing business. Banigan, as a major
importer and manufacturer of crude rubber, began to withhold payments to shippers who delivered allegedly contaminated product. In
June 1887, a New York supplier sent 300,000 pounds of variously
graded rubber to Woonsocket through a brokerage firm. Banigan sent
the first batch back, complaining of impurities: water logging, shrink-
age, over-aging, mud, and even the "doctoring" of the product. The
shipper accepted the returned lot in order "to appease" Banigan while
denying any imperfections. Banigan then refused to pay full price for
the rest of the shipment, according to the plaintiffs, "without making
any further examination." However, he would agree to take the order

at a reduced price of 5 cents less a pound. The agent placed an attachment on the Woonsocket Rubber Company, sued, but then offered to go to arbitration. Banigan accepted the terms and the two sides chose representatives, the maturation of dueling in a civilized society. Ironically, the labor movement also sought this kind of mediation in disputes with captains of industry and smaller employers, but with little success until the New Deal imposed such procedures in the 1930s.[9]

The two weeks of testimony in the case provided a detailed look into the cutthroat competition at all ends of the rubber footwear industry. The hearing may have been the first "growing out of transactions in this commodity."[10] Both sides had influential legal representation, with Banigan employing Charles E. Gorman, the Irish-American civil rights activist from Rhode Island and the state's first Irish-Catholic attorney. The cargo in question arrived in New York from Brazil in September 1886 and February 1887 and was stored in a Brooklyn warehouse until the delivery in June. The shipper admitted that some moisture may have formed during the winter months but said the product was otherwise dry and clean. Yet another competitor, Henry Hotchkiss of the Candee Rubber Company in New Haven, Connecticut, testified that he took 350,000 pounds from the same shipper at about the same time and encountered no problems. He added that Banigan was a "doubledealer."[11] Banigan admitted that he had used the second delivery of rubber within a few weeks of its arrival despite any shortcomings.

Banigan's evidence was shaky and unsubstantiated, at least in this case. The shipper's counsel attacked the Rhode Island manufacturer and his business practices in vicious fashion. The lawyer called the case important "whether a manufacturer, merely because he is rich and merely because he has in other instances succeeded in extorting allowances from people from whom he has made purchases" or "whether such a man shall be approbated and encouraged when he not only attempts, like a highwayman, to extort an unjust allowance, but seeks to do it by vilifying from the start,—men who are as much above him morally as the sun is above a lamp-post."[12] Furthermore, the attorney accused Banigan of repeatedly using the ruse in order to get a discount, "just as a dog returns to its vomit."[13] Banigan lost the case and got barbecued in the deliberations. He paid the shipper $229,184.04 including interest for the rubber and another $7,125.30

for arbitration expenses. He also suffered not only the insults but the blowing of the cover of his lucrative scheme of labeling all the imported substance as contaminated in order to lower prices. But he kept the rest of the rubber delivery and probably broke even.[14]

Banigan's grievances about the condition of crude rubber delivery may have contained some truth at times, but the case also highlighted his ruthlessness in a cutthroat business world during the Gilded Age. He listened to the barbs of the plaintiff's attorney, stinging indictments not heard since the strike a few years earlier. Most of these charges stayed buried in the arbitration report, though they probably made the rounds in the world of rubber production as uncomplimentary scuttlebutt. Banigan, during his tenure as president of the United States Rubber Company in the early 1890s, still carped about the "water and mud" in the frequently debased product. In one instance he publicly displayed a "doctored" piece of rubber. He hated to admit defeat or allow bygones to pass. However, Henry C. Pearson, editor of the *India Rubber World,* concurred with Banigan that at times rubber could be adulterated and made heavier through various subterfuges. Pearson also claimed that pilfering of the product on the New York docks turned the stolen rubber into legal tender at some local barrooms there.[15]

The National Cash Register Company

A similar accusation in another field highlighted Banigan's methodology in business before the construction of the ultramodern Alice Mill in Woonsocket in 1890 and the formation of the United States Rubber Company in 1892. By the time of his death in 1898 he sat on the board of directors of a dozen major firms. Although some of these positions were honorary, in several instances Banigan wielded tremendous financial influence. John H. Patterson, the founder of National Cash Register Company in Dayton, Ohio, and a practitioner of progressive labor-management relations, needed an influx of capital investment. Another financier connected him with Joseph Banigan. In December 1895 Patterson borrowed $10,000 for four months at 7 percent interest, pledging some of his customer's notes as collateral. Patterson's biographer, Samuel Crowther, the perceptive author of innumerable books on American finance and financiers, claimed that

Banigan lent such a small sum on purpose. Patterson, not very competent at the capitalization side of the business, would be back for more, which Banigan would parlay into a piece of the promising enterprise (i.e., equity participation). Patterson repaid the note on time, but by the middle of 1896 he required more capital for expansion. In May 1896 Banigan lent the NCR $200,000 with unusual contractual arrangements. A 7 percent rate included the pledge of the firm's receivables up to $630,000 as collateral and a clause designating Banigan as the sole lender to the firm, a form of legalized usury.[16]

Banigan had essentially created a pyramid scheme. Patterson sought the advice of H. Theobold Jr., a financier in Providence who apparently knew of Banigan's methods.

> The less we owe Mr. Banigan by paying off the yearly sum of $50,000.00 as stated in the contract, the greater the collateral security will become so that eventually we may hold herein the vault a million and a half of notes all pledged to Mr. Banigan for, say, a $100,000.00 debt; and still at such times as we would be unable to borrow any money for any purpose except through Mr. Banigan at 8 per cent with all notes then on hand pledged as collateral security.[17]

Banigan set an ironclad contract that procured a seat on the board and a seemingly everlasting 8 percent interest rate on a note almost impossible to pay off without selling the whole enterprise. The only thing that saved Patterson and the NCR from a Banigan takeover was the rubber king's unexpected death in 1898. Banigan's executor allowed an expensive alternative to the original contract through the issuing of stock for which the Banigan estate charged a carrying fee of three dollars a share.[18]

Banigan's merciless methods in most of his business dealings were fairly normal for the Gilded Age. The Darwinian principles that infected most captains of industry burned even stronger in Banigan, who knew something about survival of the fittest after escaping the Famine, Providence's Fifth Ward, and child labor at the New England Screw Company. His philanthropic activities paralleled those of other industrial moguls who contributed large sums to charity, education, and the arts in an effort to offset their ruthless reputations in business. Religious donations to churches and causes played a prominent role in the period's Protestant Social Gospel. Neither Banigan's financial brutality nor that of his Yankee colleagues often found its way into print, by friend or foe, until the muckraking era beginning at the end of the 1890s. And, like his Yankee competitors, Banigan occasionally combined a sound financial investment with a populist cause. When the novice Providence Cable Tramway sought to break the Union Railroad's streetcar monopoly in Providence in 1890, most investors balked at the unusual technology and the daunting task of taking on the unpopular but well-heeled, and established carrier. Banigan bankrolled the risky venture and it became the only cable operation ever in New England, traveling up College Hill by Brown University and serving his East Side neighborhood including St Maria's Home. Several years later the trolley cartel purchased the upstart system at a substantial price. Banigan made a profit, as usual, upgraded real estate with mass transportation, and won the plaudits of the public against the haughty streetcar company.[19]

Brazil

Banigan continually sought futuristic goals while investigating ways to improve yesterday's projects and processes. He inherently understood that the key to the regulation of rubber production was intertwined with control of the crude supply from Brazil. The rubber industry became increasingly important to corporate interests in the United States as the substance became a crucial component outside the footwear business. The application of rubber parts to ever more sophisticated machinery, medical implements, and the soon-to-explode bicycle and automobile sector, increased demand tremendously. Boot and shoe production, which had always dominated the rubber market, would account for only about 50 percent of the raw material consumed by 1890: $22,700,000 out of a total of $42,854,000. The United States government recognized this activity and established a consulate in the Brazilian state of Para at the Atlantic port city of Belem, to facilitate the importation of crude rubber in the very heart of its collection and exportation. Rubber importers influenced consuls and their staff through political connections in the United States government, sometimes in illicit involvement in foreign affairs. Charles Flint, the architect of the U.S. Rubber Company and a monopoly builder of the first order in other areas as well, wrote to the American Consul General Joseph Kerby in Brazil: "I note that you will be glad to receive my suggestions to a Vice-Consul and in response I would nominate Mr. W. B. Norton."[20] Flint, Banigan's prime competitor in the rubber import business, operated out of New York City with the legendary W. R. Grace, an Irish-born shipping magnate and two-term mayor of New York City, who used political pressure with President Grover Cleveland. Kerby wrote in the margins of the letter: "Mr. Flint is the principle [sic] American exporter from this port and whose recommendations will have much weight on acct. of his close relations with both Mr. [secretary of state, James G.] Blaine and the Brazilian Govt."[21] Banigan and Flint wrested half the rubber import industry from British domination after 1886 and eventually took control from Portugal, Brazil's mother country. However, they often paid more than the going rate. A gauntlet of different commercial handlers and myriad Brazilian taxes further inflated the price. Savvy deals could cut costs, and the price gyrated wildly according to supply, demand, and speculation. Prices actually fell for good at the same time the rubber interests were trying to sell the concept of a monopoly to the American public as a stabilizing deal for consumers.[22]

Banigan never enjoyed enough political capital to rival Flint in that arena, but he focused so intently on the business that he undermined his competitor on several occasions with his hands-on approach to matters. He traveled to Brazil as early as 1882 and eventually established his own import house with offices in two Amazonian cities. He employed the powerful and influential firm headed by LaRocque da Costa and Company, using English letters of credit.[23] At first he observed, incognito, the harvesting of crude rubber and the financial machinations that accompanied these operations. He apparently secured a good deal on a supply of rubber, and the local merchants told him they would sell the remainder to "the distant Banigan" at an exorbitant fee! Many sources identified Banigan as the largest importer of rubber to the United States in the 1890s despite the wily deals of Charles Flint, who would later court Banigan and the Woonsocket Rubber Company to join the monopoly. Banigan was probably the only footwear producer actually to visit the South American country, a remote destination at the time. He bragged that "when I was in Brazil I was surprised to find Woonsocket shoes."[24] One of the local newspapers stated that he "became as familiar with the topography of the rubber country as he was with that of Woonsocket or Millville."[25] Banigan gave a detailed analysis and first-person account of conditions in Brazil to the Boston Boot and Shoe Club in 1895. His power of description still enthralls the reader almost as a travelogue and journey with the native rubber gatherers.[26] In a court case in 1898 he declared: "I have been importing rubber for a long, long time, and in very large quantities; probably the importer of the largest quantity, individually, of any man in the United States, and possibly in the world; very large quantities."[27] He decorated the company's office with artifacts and rubber implements actually used in Brazil as well as with a collection of other native items that turned the room partially into an anthropological museum. The model of the steamship *Metis* along with a portrait of Brigham Young must have stood out among the eclectic and primitive pieces.[28]

In early 1890 Banigan announced that he had made arrangements to import crude rubber from the Amazon directly to his Woonsocket plant via the port of Providence, allowing him to control "the raw rubber output of the world."[29] The bold decision to bypass New York rubber merchants roiled the waters of the industry as the steamer *Brazil* stopped first in Providence to unload before heading on to the more traditional dockyards of Brooklyn. A delivery in early 1891 required forty railroad boxcars to transport the material to Woonsocket. Even more spectacular was Banigan's unfulfilled challenge to Rhode Island businessmen to find the requisite goods to fill the empty cargo space on return trips to Para, Brazil.[30] Banigan shaved an impressive one-third of the carrying charges by skipping the expensive New York harbor facilities, but the possibilities of outfitting a ship to South America with various Rhode Island goods caused an even greater froth among local entrepreneurs.[31] "The coming of a steamship direct from Para with a rubber cargo, was what caused the eyes of the shipping men to open wide," marveled the local board of commerce.[32] Actions like these impressed his Yankee counterparts in business, raising Banigan's stature among an important but unfriendly group. Over time these businessmen warmed to the Banigan touch. The initial delivery of crude to Providence weighed 134 tons. Rumors circulated at the time of the first transfer that Banigan, with his sudden popularity among Yankee financiers, might run for the United States Senate, a selection still determined in the General Assembly, not by a popular election.[33] The *India Rubber World* listed the amount of rubber imports to major U.S. cities, with Providence showing up for the first time in 1888.[34]

Woonsocket Versus Millville

If Banigan pretended to work in concert with competitors and cartel builders, the truth emerged at the end of 1888 when word leaked out that he sought a tax break from Woonsocket to construct a new footwear facility in the city to replace the original 1860s piecemeal operation there. The planned building would accommodate 3,000 rubber shoemakers, the largest such facility in the world. The construction of such a behemoth sent shivers among those seeking to create a monopoly as Banigan's plans suggested he was prepared to go it alone. Furthermore, he whipsawed Woonsocket against Millville again as in 1882, vowing to build in whichever municipality provided a five-year tax break. The Woonsocket Business Men's Association pleaded with Banigan to stay in Woonsocket. He spoke again at their annual meeting in January 1889. A special committee had already met with the rubber king and they must have felt as vulnerable as the strikers in 1885. The group agreed to endorse his demands: a five-year tax exemption on the new mill, a similar break on any subsequent improvements, and reimbursement of the appraised difference between real estate in Woonsocket and in Millville. Lastly, the city would pay for any infrastructure developments. The committee of six businessmen recommended to their one hundred colleagues that the association go on record endorsing the clauses and encouraging the city council to do the same. "We believe this to be wise," they concluded, "as it will rebound to the future good of our city; judicious, as it is not injurious to anyone, but beneficial to all, for if the works go to Millville we lose what we already possess; practical, as it will be a good initial step for our new city [incorporated in 1888], and when once these works are perfected and running other industries will be encouraged to locate here and share the many advantages of our new and prosperous city." Woonsocket had to introduce a bill in the state legislature for the exemption.[35]

Banigan's speech to the membership had the ring of a contemporary developer. Apparently not comfortable as a public speaker, he made a Freudian slip. "In 1882 we were unfortunate enough to feel obliged to remove in part [referring to the erection of the boot factory in Millville] from your nicely located town. Our town, I mean, for I am a taxpayer here, and please do not consider I am altogether selfish."[36] Banigan complained about land values in the city, the distance to railroad facilities, and the costs involved in cartage. He also reminded the audience that the facility, patched together as it was, employed 800 workers and produced 8,000 pairs of shoes daily. Banigan got carried away praising the distribution of the Woonsocket brand on every boot and shoe he produced. "Thus the name of the city is sent to every state in the country," he bragged, and "to England, Scotland, Germany, Russia, Australia, China and Japan, for we have a market in all these places." He assured the association that any costs would be easily recouped from the increased payroll. He openly displayed a preference for the city. "Woonsocket has the advantages of a good location, good and numerous churches and schools and a good city government." Perhaps remembering the rowdy nature of the 1885 strike in Millville, he commented, to its detriment, "in Massachusetts they have a good many grogshops."[37]

Millville, already home to the greatest plant in the world, also salivated at the possibility of adding a counterpart from the shoe world, however remote the possibility. Banigan owned sufficient land next to the Millville plant to construct a new facility with the added convenience of an adjacent railroad. The town sponsored rallies and, at a taxation vote in Blackstone, "every voter in Millville was present."[38] Ironically, John Conway, a founder of the original Knights of Labor assembly at Millville during the strike, organized support for the new hometown factory. He now ran a small grocery business in partnership with Joseph Hines, the president of the Order whose dismissal had triggered the walkout. He seemed sincere in his efforts to persuade Banigan to let bygones be bygones, perhaps looking at the larger economic impact, seeking reemployment with the company, or just envisioning more customers for his own venture. Banigan spoke at the gathering and, while admitting that Woonsocket was the birthplace of the firm, he promised to keep an open mind.[39] Although the tax assessors agreed to match any concession from Woonsocket, at least one manufacturer in town threatened to sue over the issue of preferential treatment. Banigan later upped the ante with a highly unlikely threat to sell the Millville facility and consolidate all operations in Woonsocket in a mammoth, centralized plant. Meanwhile the city of Pawtucket entered the fray offering twenty acres of land to situate the enterprise at the origin of the Industrial Revolution. Woonsocket, however, felt it held the trump card by already housing the concern but took no chances and proffered a ten-year deal that would limit the valuation of the new building to no more than $100,000 a year. Banigan, the ultimate card player, raised the stakes even further. His final offer to Woonsocket demanded a $100,000 ceiling on taxes for fifteen years on any of his property there and a full municipal subsidy for a new road and bridge to the proposed location. "If the city of Woonsocket cannot afford to do this, we shall be obliged in our own interests to build at Millville." Banigan had already crossed the Rubicon or, in this case, the Blackstone River, and bought Buffum Island for the new site in Woonsocket in March 1889. The rest was all theater. There was no philanthropy at the point of production during the strike in 1885, nor would there be any hometown charity now where the rubber met the new road to his palatial mill. Banigan chose Woonsocket on his terms.[40]

The Death of a Matriarch

Almost a year before the stunning factory opened, Joseph Banigan's mother, Alice, died at the age of seventy-eight in October 1889. The matriarch of the family had outlived her husband, Bernard, by a decade. Dozens of priests from many of the state's parishes attended the obsequies at the Cathedral, of Sts. Peter and Paul, her place of worship for forty years. Her brother's son, the Reverend Patrick H. Finnegan, assigned to a parish in Connecticut, celebrated the solemn requiem mass, providing a connection to the seemingly ancient times on the Shirley Estate in Ireland. Father McCabe from Woonsocket, Father Kittredge from Millville, and Father Meehan from Natick, home of the textile bleachers from Ballybay, County Monaghan, Ireland, who preceded the Famine generation to Rhode Island, joined the mourners there as well. The Reverend Robert J. Sullivan, who had officiated at the opening of the Little Sisters of the Poor, paid tribute to Alice Banigan as an actively charitable woman herself, perhaps the fountainhead for her son's own philanthropy. Father Sullivan ended the sermon by saying, "Let you, dear friends, and above all, you who were the recipients of her kindness, join with her relatives in remembering her for many years to come in your prayers, and join now with me from your inmost hearts in saying, May God have mercy upon her soul." Banigan's success and his mother's longevity assured at least some public note of her death.[41]

The state's Catholic newspaper wrote that "the funeral cortege was very large, many of the oldest and most prominent Catholics of the city and State being represented."[42] Most of the pall bearers worked in management at the Woonsocket Rubber Company. Alice Banigan was buried alongside her husband at "the old Catholic cemetery," belonging to St. Patrick's Church on Douglas Avenue in Providence. Both would be laid to rest again in the mortuary chapel Banigan was about to build for the family in Pawtucket. Family and friends remembered her as "Grandma Banigan" who presided at Thanksgiving Day festivities at her home. The last five of those holidays featured four generations of the family, Joseph's children having their own broods now. She had held court at 59 Dean Street, not far from the old Irish Fifth Ward.[43]

The Alice Mill

The four-story Alice Mill, named after Banigan's recently deceased mother, opened in Woonsocket on 23 October 1890, with a sumptuous banquet and dedication. Rhode Island's premier caterers, Gelb and Norton, provided an eye-popping feast. Each setting contained a pair of miniature rubber boots embossed with the company's name. Banigan invited five hundred guests. The Providence and Worcester Railroad added three private cars for the trip from Providence, just for the out-of-state visitors. The train spur took them directly to the facility, one of Banigan's requirements for the site. They entered the second floor of the great building, the largest rubber footwear facility in the world with a capacity for 30,000 pairs of shoes a day. A 1,500 horsepower engine now ran the factory compared to the fifteen-horsepower one used when Banigan started the operation in 1866. Each of the four floors measured seven acres. The company expected employment to crest at three thousand operatives who could turn out a pair of shoes in half the usual time owing to automation. Decorators festooned the building with streamers, bunting, and flowers. A large

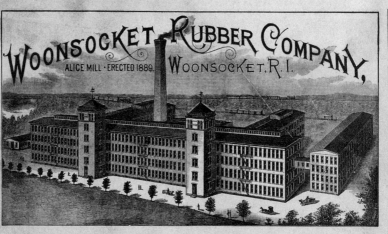

THIS MILL IS NOW IN OPERATION

Manufacturing a Full Line of

WOONSOCKET RUBBER SHOES.

Banigan established the Alice Mill in Woonsocket in 1889, the premier shoe production establishment in the world. The ultra-modern plant almost stalled the formation of the United States Rubber Company in 1892 because none of the cartel's members had a comparable facility. (Boot and Shoe Recorder, 8 October 1890. Author's collection)

American flag hung from a back wall with two smaller Irish flags on either side, a direct reminder to the guests that this was not a typical Yankee enterprise despite the integrated guest list, which the local newspaper published. The group did not include any recognizable rubber competitors. Some visitors had a nostalgic pedigree: Lyman Cook, the original backer of the Woonsocket Rubber Company, now in his eighties; John Holt, Banigan's beloved father-in-law and colleague, already retired; and Margaret L. Banigan, the rubber king's missing aunt, sought and found through an ad in the *Boston Pilot* in the 1850s. She lived in Atlanta, Georgia, where Banigan later purchased an entire block of expensive commercial property.[44]

Woonsocket enjoyed the greatest celebration in its industrial history. The speaking program accentuated the most memorable day in Banigan's business career. Daniel Pond, the mayor of Woonsocket, gushed at the putative employment of several thousand and the estimated $50,000 monthly payroll. He complimented Banigan as "a king in his business." The governor of the state, "Honest" John Davis, the only Democratic chief executive in Rhode Island during the Gilded Age, toasted Banigan in a biblical panegyric: "Seest thou a man diligent in his business, he shall stand before kings." The construction superintendent, Patrick J. Conley, proudly recited the establishment's cutting-edge features: Edison electric lighting, a modern heating system, and even air conditioning for the summer months. The rubber king rose, the last speaker of the day, to provide a secular benediction. He said that "he had been talking for the company for twenty-four years, and it was hard to say anything new." He bragged of the nine million pounds of crude rubber consumed in his factories yearly and could not resist the joke that it was not all pure rubber, a barb that referred to the trial about watered-down and contaminated product that garnered great laughter from the knowledgeable crowd. When he finished, the audience responded with "thunders of applause" and the band played "Hail to the Chief."[45]

The distance from the potato patch in Lisirrill was beyond measurement. Yet Banigan's future prospects loomed even brighter despite the fact that key rubber moguls, absent from the festivities, were already planning a rubber monopoly without him. Banigan, with the completed Alice Mill, could flout the company's independence from any industrial trust while placing a different kind of trust in the political and commercial sophistication of the buying public. Like Henry Ford a generation later, Banigan posed as the champion of the consumer against the corporate octopus.

The Knights of Labor Redux

On the other side of the great labor-management divide during the Gilded Age, the local Knights of Labor prepared as assiduously for the altered conditions in the rubber industry as did the principal capitalists forming the cartel. After the 1885 strikes the local Order took under its wing fourteen blacklisted members. It placed a notice in the *Woonsocket Evening Reporter* guaranteeing to settle any bills accumulated by the outcasts during the strike. Over the next couple of years the Knights held a series of social and organizational events in the area to raise funds and solidify the union politically. Irish-American activism continued unabated and often dovetailed with union enterprise. In January 1886 young members from Millville trekked to Boston to obtain citizenship papers, 180 making the nearly forty-mile trip in one day alone. A few months later "men from the rubber works who have been naturalized recently" registered to vote in Millville. Labor activists formed at least a temporary political party there while two hundred Woonsocket Knights marched in the Order's parade to champion a change in the Rhode Island state constitution to allow workers to vote without property qualifications, a reform that came in 1888. Furthermore, the Millville Knights formed another assembly, LA 7226, a mixed union of unskilled and semiskilled operatives at the rubber facility to complement the original bootmakers and cutters group. This event, in the words of the Order, provided "cellar to attic coverage," undercutting Banigan's charge that skilled workers selfishly abandoned their comrades when they first left work. Organizing the unskilled also answered in the affirmative, at least in this case, one of the pressing historiographical questions as to whether the union's elite constituents encouraged their less qualified co-workers to unionize. The *Woonsocket Patriot* characterized the new constituents of the Knights "as some of the most popular young men of the town," who seemingly provided a solid base for continued growth.[46]

Joseph Hines, the bootmaker and president of the Knights whose firing helped precipitate the second strike, worked for a while at the

Franklin Rubber Company. He was single and lived with his widowed mother in Banigan Village until they were both evicted during the walkout. They moved to Woonsocket, but Hines eventually returned to Millville. The town, perhaps at Banigan's urging, had stripped Hines of his voting rights until he complained after reestablishing his residency. The authorities quickly restored his privileges. In March 1886 the Millville town elections were "the scene of a larger attendance, heavier vote, and more excitement than the oldest inhabitant can recollect." The activist Knights, which supported local Democrats in this joint venture into labor political activity, won almost all contested positions. Furthermore, the Order teamed with the Catholic Church to endorse a local ordinance prohibiting liquor sales, prompting this editorial comment from the *Woonsocket Patriot:* "when good citizens, the Knights of Labor and the Catholic Church, unite in a good cause like this, . . . some lasting benefit must come from their action." A few months later, the Democrats, with the help of the Order, won all seven council seats in Millville.[47]

Largesse to the locked out Knights continued as District 30 parceled out $600 in late April, but the delegation refused by a vote of 99 to 67 to allocate weekly stipends to each of the remaining blackballed members, preferring to pay incidental living expenses. Perhaps the state governing body of the Knights feared another obligation that threatened to outlast even the length of the strike. The Millville veterans also became part of a larger and continuing effort to organize the entire rubber footwear industry in the East under the aegis of the Knights. The strike at the Woonsocket Rubber Company precipitated the issue of a larger, industrial-style union for rubber workers. In fact, Millville operatives assisted co-workers at the Bourn family's old National Rubber Company in Bristol during a walkout there not long after the one in Millville. Samuel P. Colt, Republican politico who eventually ran the United States Rubber Company after Banigan, now controlled that firm.[48]

In March 1886, union delegates from the major rubber companies in the region met in secret session for two days and formed the United States Rubber Employes [*sic*] Protective Association. The 57 attendees allegedly represented 53,000 rubber workers, but the proceedings stayed private. A month later Massachusetts District Assembly 30 of the Knights of Labor heard arguments for and against a resolution to endorse a "National Trades Assembly of Rubber Workers," introduced by Edward S. Blaine of LA 3982 in Malden at the state convention in Boston in April 1886. While the rubber LAs stuck together, Boston labor luminaries like Albert Carlton, George McNeill, and Charles Litchman, all of whom had visited the Millville strikers, spoke against the measure. However, their opposition to the resolution had more to do with internal ideological disputes that were pulling the Knights in two separate directions at the time, than with the merits of the proposition. McNeill, for example, believed that any retreat to a craft-based organization undercut the solidarity of the new working class and abandoned the inclusive nature of the Knights for the narrow, exclusive craft terrain of the new American Federation of Labor. He wrote that "every argument in favor of forming trade districts out of our mixed districts is an argument against the methods and principles of our Order. It is the attempt of a revolution to go backward." At the next quarterly meeting of DA 30, held in July in Worcester, five rubber locals in three towns (Millville, Malden, and Franklin) petitioned the organization to withdraw "in order that we may organize under a charter from the General Assembly, Knights of Labor, as a National Trade Assembly of Rubber Workers, and thereby be enabled to more effectually carry out the objects of our noble Order as applied to the rubber industry." The motion, submitted by Millville LA 3967, was rejected. Woonsocket LA 4049 similarly petitioned the Rhode Island District Assembly in the exact same language, but the motion was tabled and then, in October, withdrawn. Undeterred, the rubber workers held another confidential conference in Connecticut the following year. Ironically, leather boot and shoe workers successfully formed their own trade assembly in 1885, provoking a fight with a couple of state DAs.[49]

The internal solidarity of the footwear workers could not overcome the internecine battles raging within the Knights over the very question of a trade union orbit within the Order. Frank Foster, the former editor of the Knights' newspaper in Haverhill, Massachusetts, now published the Boston *Labor Leader.* He reported after the local convention that Edward Blaine was attempting to establish a cooperative factory, a hallmark of the Order, which included a textile coop and union-run restaurant in Rhode Island. A month later Foster visited Blaine at Malden and declared that "the rubber workers have been trying to take advantage of their constitutional privileges and form a national trade assembly, but in some instances have met with objection." Blaine continued to publicize his advocacy for rubber solidarity, speaking at the Knights' national convention in Richmond, Virginia, in October 1886, giving "the trade unionist idea a great lift in the order." Blaine envisioned a confederation of national trade assemblies that would control the destinies of particular crafts or industries. The *Labor Leader* minimized the differences between the Order and "the great national trade unions" of the American Federation of Labor as one of preambles, initiation, and rituals. However, the philosophical rift was too great for the two labor organizations: inclusion vs. exclusion of different workers, idealism vs. pragmatism, and differing views of the roles of skill, nationality, race, and gender. For many Knights this seemed dangerous ground and represented an abandonment of the wider objective of comprehensively reforming the industrial order in order to settle for the more limited goals of collective bargaining. Although some trade districts in the Order originated before 1880, they did it on a haphazard and shifting basis before the formation of the AFL and the hardening of positions. However, as late as 1885, District 99 in Rhode Island created a state Central Labor Union to house craft unions exclusively.[50]

At the famous 1886 national assembly in Richmond, the heart of the old confederacy, DA 144 of Connecticut introduced a resolution that craft workers "should organize into National or International Trades' Assemblies, whereby all matters directly interesting these trades may be under the jurisdiction of members of such trade, who are better qualified to settle difficulties in their own calling than those who have no knowledge of the same." This attitude was in direct competition with the Knights' traditional practice of mixing assemblies with representatives from different occupations where, the thinking went, an outside opinion would provide a thoughtful perspective and create a stronger vehicle for general reform and industrial equality. A similar measure was introduced by the other Connecticut DA, number 95. The convention's Committee on Law introduced a seven-section constitutional amendment governing the formation of national trade groups. Edward S. Blaine, one of the Massachusetts delegates, made a substitute motion that would allow five or more local assemblies with at least one thousand members in the same trade to apply for a charter, providing they had paid all dues and assessments owed to any state body, a bone of contention in some of the powerful districts, which feared the loss of so many *per capita* dues. In Blaine's proposal, the national trade assemblies would be autonomous within their own jurisdiction but still come under the auspices of the national organization. Several friendly amendments were attached, and the language was adopted unanimously by the delegates. Although there is no list of rubber workers categorized by ethnic origin, the Irish at the Woonsocket Rubber Company, and the many who left there and branched out to other similar enterprises for one reason or another, probably gave that industry a Hibernian tint in New England, as it also had within the Order.[51]

In the rubber industry there seemed to be no underlying desire or secret agenda among workers to abandon the Knights. Union activists in the footwear business wanted to fine tune and expand an organizational structure already in place within the Order to compete with significant structural changes being considered by management in the context of expanding regional, national, and even international economies as demonstrated by the precursors to the cartel in 1892. The local rubber assemblies in the Knights, if anything, displayed a strong allegiance to the Order because of the organizational and financial assistance provided during the 1885 strike wave. Knights in the rubber industry exhibited a keen awareness of the approaching transition within the trade and of the larger merger movement within business circles in general. The Knights did little to follow through on the mandates of the convention in 1886 because of sustained hostility by the national leadership, state assemblies, and a proposed treaty between the KOL and the AFL to leave trade union activity largely in the hands of the latter. A few years later the Order changed direction to some degree in granting national trade charters, but more as a competitive wedge against the American Federation of Labor once the relationship between the two groups soured for good.[52]

The internal break in the Knights came on 21 March 1887 when ten rubber locals, including the Millville and Woonsocket locals, two New York assemblies, and the oldest rubber union still in existence in New Brunswick, New Jersey, applied for a "Rubber-Workers' National Trade Assembly of America." When the general executive board did not respond to the charter request within two weeks, New Brunswick,

LA 3354, sent an ultimatum either to issue a charter within a week or the rubber assemblies would leave the Knights and "form open unions." The New Jersey local wrote that it already had the support of the AFL's local Essex County Trades' Assembly and the New York Central Labor Union. Charles Litchman, who had been involved with the Millville strike and was now national secretary, claimed that the LAs had not secured permission from state District Assemblies to properly withdraw under the constitution and form a national trade assembly, as he "repeatedly" reminded the assemblies. The Knights dispatched New Jersey's district master workman to visit the recalcitrant union, but he was snubbed and suspended the chapter. The renegade assembly then notified the press that ten thousand rubber workers would abandon the Order, but Litchman ruefully noted that the LA had only seventeen members in good standing. The *Labor Leader* newspaper, now thoroughly trade union oriented, remarked disingenuously that the secession by New Jersey was triggered "on account of the way the Knights allow the scabs in Woonsocket to compete with them." The rupture was more symbolic than surprising, as other local assemblies broke away when denied a national charter in other industries as well. In some instances the American Federation of Labor contained no corresponding craft union to take in these orphan locals from the Knights of Labor.[53]

Before the Knights' convention met in Minneapolis in 1887, the Millville assemblies hosted a Labor Day celebration: "Such a multitude was never seen in our busy village as that day, every train bringing hundreds. The big attraction was the Knights of Labor picnic." The following April Edward Blaine and Frank Foster addressed a meeting of the local unions and probably advocated trade union status. On Labor Day, 1888, the same Millville assemblies held another picnic. Blaine, the crusading secretary of District 30 and supporter of a national charter, spoke, as did James H. Mellen, the enthusiastic editor of the *Worcester Daily Times* and now the Knights' master workman of Worcester County. Mellen blasted the increasing use of piecework in the rubber industry and the tendency toward monopolization. He reiterated his support for the strikers of 1885. Now that he was a representative in the Massachusetts legislature, he allegedly introduced the first weekly payment bill in the country.[54]

But the end was close at hand for LA 3967 in Millville. In March

1890, "The rubber workers' assembly, Knights of Labor, . . . surrendered its charter and disbanded," according to the *Woonsocket Patriot,* "It was once a powerful labor organization." LA 2297 had been the pioneering local in the Order's stable of rubber footwear producers (both boots and shoes) in 1882 at the original plant of the Woonsocket Rubber Company. The axis of labor power shifted dramatically to the heavily male, highly skilled bootmakers in Millville when the new factory began operations there later that same year. During the strikes in 1885 the cutters in Millville, who probably initiated the first union in Woonsocket as well, organized a separate assembly, LA 3967, now terminated, to do battle with Banigan. The older assembly in Woonsocket, deprived of the skilled male craft workers, morphed into a "Ladies' Assembly," LA 4049, and played a small role on the periphery of the walkouts (during the first recorded strike at the Woonsocket Rubber Company in 1875, women played a more militant role than their male counterparts). With the completion of the Alice Mill in 1890, all shoe production was consolidated at the new facility. The workplace there became even more female in the highly automated plant. The deskilled women fended for themselves as the male Millville union ended its existence as an affiliate of the Knights. At the same time, Edward L. Gannon, an influential leader of the Rhode Island Knights of Labor who represented Burrillville in the General Assembly, spoke about organization in a public venue under the auspices of the same Ladies Assembly. A month later the female shoemakers held an invitation only dance with forty couples, attempting perhaps to resurrect the LA. However, four years later, in 1894, the assembly at the Alice Mill was described as almost moribund, "hardly anyone has been aware of its existence." The female operatives were "quiet and careful in their movements, and most of them are adverse to public demonstrations and even keep the names of the officers to themselves, as far as possible, never at any time giving election lists for publication." They only represented 130 of the 1,500 operatives at the plant and reportedly acted as secretively as in the earlier days of the Order. Interestingly, in 1894 the Rhode Island Knights unsuccessfully lobbied for Nellie Marra, a shoemaker at the Woonsocket firm and the Order's statistician, for the post of state factory inspector.[55]

In Millville, the skilled bootmakers congregated in LA 3967 and, in 1886, formed the semiskilled LA 7226 to encompass other operatives there. The artisans continued to escape the clutches of a fully mechanized operation. The size of the large and unwieldy boots evaded the assembly line process for a while longer. The skilled and semiskilled male workers continued to enjoy a degree of autonomy during the twilight of the Knights, but the handwriting was on the wall when five sole-cutting machines arrived. By that time women already ran the packing room in Millville. Management callously exploited female operatives in the mechanized plant as a weapon in deskilling the larger workforce while the Knights' earlier interest in the gender frontier fell by the wayside. As the Order disintegrated, the progressive experiments in diversity, cooperative ventures, and day care nurseries became the first victims of a falling membership and diminished finances.[56]

The rubber workers in the Knights of Labor, as well as other craftsmen, had gotten a whiff of the coming age of monopoly. Their struggle in 1885 to organize all the rubber footwear producers in the industry, if successful, would have prepared them well for the introduction of the United States Rubber Company seven years later. After the New England walkouts, they pressed the parent General Assembly of the Knights of Labor to wake up to the new realities of industrial life. They peppered officials to make the organizational changes to level the playing field so that organized rubber footwear workers throughout the country could engage the forthcoming cartel with some degree of solidarity.

When Edward Blaine wrote a report on the history and prospects of labor organizing in the rubber industry for the Massachusetts AFL in 1890, the *India Rubber World,* one of the trade journals, reprinted part of it. The editors introduced the piece by warning: "That growing association known as the American Federation of Labor, in five years' time has formed thousands of 'Unions' in all parts of the United States, and representing almost every trade that can be mentioned. In the recent report of the Massachusetts State Branch is to be found much of interest to the rubber man." The tireless Blaine gleefully described a reawakening of rubber workers about 1890, establishing AFL chapters of the trade at the state and national level. There were now eight federal rubber locals in Massachusetts, which meant the branches enjoyed direct affiliation to the Federation prior to the establishment of a national union for rubber operatives. The editors of *India Rubber World* investigated the dues payments to the Massachusetts Federation of Labor to identify five of the eight local sections. Although Millville was not mentioned, the unidentified AFL Federal Local 4088 was organized on 25 March 1890, about the same time the Millville Knights turned in their charter. So pronounced was the clamor for craft status in the region's rubber shops in 1891 that Samuel Gompers named the president of the local at the Para Rubber Company in South Framingham as the general organizer for rubber unions in the AFL. A year after being elected president of the U.S. Rubber Company, Banigan closed the Para plant as inefficient, once again decapitating the union leadership as he had during the Millville outbreaks. The workers were without a union for two years after the fragmentation of the Knights in 1888. They suffered a 35 percent wage cut until joining the AFL.[57]

The Federation's Rubber Workers Union eventually would form a strong national body, although the production of bicycle and automobile tires would reorganize the industry by product line and geography. The Knights, despite internal bickering and organizational myopia, expired more as the result of a vicious management counteroffensive than of any inherent problems. The Order shocked and unified management into a national and state-by-state showdown that the Knights could no longer sustain. In Rhode Island, the Slater Club, a forerunner to contemporary chambers of commerce, bankrolled a local factory in a labor dispute that eventually bankrupted the Knights into surrender in 1887. The Panic of 1893 did the rest, scattering many of the remaining Knights locals into independent cocoons or else destroying them altogether. The same financial disruption prevented the American Federation of Labor from picking up the pieces immediately. In the Woonsocket-Millville area workers suffered unemployment owing to the recession. At an out-of-work gathering, a local newspaper reported that the participants "acted like men afraid of Banigan," despite rumors from the board room of the United States Rubber Company that the old boss would look out for his former help. The demise of the Knights, however, tipped the balance of industrial power back to the manufacturers, at least temporarily. When the Belgian Yarn Company decided to establish the River Spinning Company in Woonsocket in 1891, the owners claimed "the city is freer from all kinds of labor troubles than any place visited by them."[58]

As conditions and structural change punctuated the world of rubber production, time also caught up with many of the protagonists at the Woonsocket Rubber Company. Father McCabe, after thirty-six years in Woonsocket, died just before Christmas 1893, causing an unprecedented outpouring of grief. "No scene like it was ever witnessed in this city." Bishop Harkins led an incredible battalion of 160 priests, 12,000 mourners, and every Irish-American organization in the area at the funeral. A local newspaper sympathized that "the dearest wish of his heart was to see that country free. He was a generous contributor to every subscription made for the good old cause of poor old Ireland." Father McCabe was buried at the front entrance of St. Charles church under a granite monument. A marble bust of the sixty-six-year-old vicar general graced the inside vestibule. Banigan attended the funeral, but the long oration never mentioned the strike controversy at the Woonsocket Rubber Company or the courageous role played by the beloved, even-handed pastor.[59]

John Holt, the popular superintendent at the Woonsocket Rubber Company and a behind-the-scenes advocate for workers, served as a pallbearer for Reverend McCabe. He died just a few years later himself in 1896, leaving most of his considerable fortune to Catholic charity. The *Providence Visitor* wrote that "his conversion to the faith was the reward of his inquiry after truth and unswerving fidelity to conscientious convictions."[60] Holt's funeral drew a large crowd as well. Father McCabe's successor at St. Charles said of Holt that "he never curtailed the wages of the poor man in order that he might become wealthy."[61] That was as close as the media ever came to mentioning anyone's role in the 1885 upheaval when Holt stood along with Father McCabe against his son-in-law. The strike was basically stricken from the public record. Business history remained, but labor's role was greatly diminished. The old guard and an old order were changing.[62]

THE UNITED STATES RUBBER COMPANY

....

> *"The People have formed a Trust with the [Woonsocket Rubber] Company and the People's trust is bound to succeed."*
> —Boot & Shoe Reporter, *October, 1889*

> *"There was a thorn in the side of the United States Rubber Company so long as Joseph Banigan remained on the outside."*
> —India Rubber World, *1 Jan. 1899*

Joseph Banigan, more than any of his rivals in the rubber industry, designed an empire from the bootstraps up. He accumulated an encyclopedic knowledge of the trade and might have been content to continually refine and expand his own enterprise. With a few more years of independence, the Woonsocket Rubber Company might have become its own cartel. His competitors, eager to join the rush to conglomeration in the 1890s, tried to pool their resources without him, but they quickly realized how indispensable the Banigan method was to the work of consolidation and that they needed his participation. The fledgling United States Rubber Company made an offer that Banigan could not refuse: a veritable king's ransom for his creation.

Banigan took the purchase price but could not run with the money. His life was so intertwined with rubber production he was psychologically unable to retreat from it. He convinced himself that he could replicate his experience with the Woonsocket venture among the other concerns. The rubber king soon discovered, however, to his great dismay, that, once a captain of industry he lacked now the capacity to create vassals to his own larger-than-life commercial monarchy. He pocketed the profit from the sale of his kingdom, but, even though he possessed the formula for progress, was unable to make his former adversaries cower. The rubber king could not countenance his new-found allies as knights any more than he had been able to accept his rubber workers as Knights of Labor. In the long run, though Banigan had defeated the operatives, he could not vanquish jealous associates. He took a fall.

Joseph Banigan participated, willingly or not, in the thrust to create a rubber cartel in the United States. At some point many large industries, especially in the 1890s, dumped the cutthroat competition that divided them and also tamped down potential profits. Although cooperation ran against the grain of their competitive instincts, these localized businessmen understood the benefits of collaborative production, monopoly control, and domination of the raw material market. V.I. Lenin's later theory that imperialism was the highest stage of capitalism resonated with those on a quest to establish a trust to control Brazilian rubber. They also appreciated the efforts of Rhode Island's senior senator, Nelson W. Aldrich, who tried to lower import duties from his perch on the U.S. Senate's finance committee.[1]

Although Banigan went along for the ride, he never fully surrendered to corporate solidarity, any more than his cohorts did, even if his enterprise served as the cartel's flagship. That concept of unity was better left for the dying Knights of Labor and Samuel Gompers' nascent American Federation of Labor. Banigan maintained just enough independence to control the fulcrum of power. He fortified the Woonsocket Rubber Company so completely that his participation or truancy in the endeavor could make or break the trust. The Alice Mill, combined with the still advanced Millville facility, meant that no alliance could topple his empire. With his technological prowess and secure supplies of raw rubber, he could linger awhile to ascertain the prospects of the combination. His public relations acumen and steely willpower could dictate terms regardless of the brewing strength of the opposition. He knew in his heart they could not succeed without him. If he decided to relent and join the business pool, his rivals would surrender on his terms.[2]

In 1889, the year he started construction of the Alice Mill, Banigan joined the latest sprint in the race to monopolization. The Rubber Boot and Shoe Manufacturers' Association, comprising ten of the fourteen key facilities in the United States, once again joined forces to combat the spiraling cost of crude rubber as demand for its use in other products pinched supplies. Banigan took part in organizational meetings in New Brunswick, Boston, and elsewhere in the late summer and early fall of 1889. This time around, the combined forces of American rubber production engaged British capitalists, Brazilian speculators (Banigan paid $40,000 in export duties to Brazil for a single shipment in early 1890), Portuguese taxes, and internal U.S. consumption. Although Banigan participated in the organization, he advertised his firm as opposed to any methods inimical to consumers: "The People have formed a Trust with the [Woonsocket Rubber] Company and the People's trust is bound to succeed."[3] The incipient cartel was not insensitive to public opinion either and blamed Brazilian and Portuguese syndicates in Para for the high price of crude rubber and the consequent increase in the cost of footwear. Banigan might play the outsider, but he met in London in July 1891 with E. S. "Deacon" Converse of the Boston Rubber Shoe Company, Samuel Colt of National Rubber in Bristol—a key player in the putative monopoly—and a representative of the Brazilian syndicate. They probably discussed impending mergers, although the industry suspected a holdout by Converse and Banigan, with the latter importing rubber into the United States for the two independents. Soon after the meeting, the Brazilian consortium stumbled when British banking interests demanded a greater interest margin, deflating the price of crude almost immediately. At the same time Banigan brought home a contract from a British buyer for one million pairs of footwear. *India Rubber World* poked fun at the conspirators in the form of a Shakespearian play, "The Tubber Rust," with Banigan as the Earl of Woonsocket, "A Huckster," who says:

"Tis truly said, my lord, in times since past some words we've had, some misunderstandings. But, with all my heart we'll deep bury them, bravely fight the common foe, low prices. True, I oft and deeply cut on prices and thus did irritate thy noble soul. O, my offense is rank, it smells to heaven! But what boots is now?"[4]

Charles R. Flint

Charles R. Flint (1850–1934), a wily and brilliant entrepreneur who possessed unusual talent to orchestrate cartels in different sectors, succeeded in several key industries including wool, electric lighting, and even starch and caramel. He operated almost behind the line of trust-busting fire, willingly sinking to a second tier of public and political identification behind the Goulds, Carnegies, and Rockefellers. He operated out of the glare of the emerging muckrakers and often kept his corporate charges in the dark, yet wielded tremendous influence. He planned the formation of the U.S. Rubber Company, but his name did not appear as an incorporator. He possessed a commercial pedigree from his family, which owned one of the largest shipping fleets in the United States. Flint initially made a fortune in munitions, banking, the sale of naval war ships to foreign governments, and by an occasional foray into global intrigue. As a partner in the established firm of W. R. Grace and Company, whose specialty was in South American commerce, he acquainted himself with the rubber industry and by 1878 began dealing in crude from Brazil while holding some stock in the production end of the industry. He kept in touch with Senator Nelson W. Aldrich, soon to be known as "the General Manager of the United States," about import duties and banking opportunities. An undated and unaddressed letter by Flint underlined his almost encrypted methodology to manipulate supply and demand and drive up the cost of rubber to his advantage:

> Now, the $1,500,000 man says buy at A get Grace, Jews, Gould, and others sell largely short for which one would use friendly Mfgrs. here and in Europe to buy—then put market up to 80 cts. Get G. J. & Co. sell more short. Put up to 85. get competitors sell still more short, then put and keep price at $1[.] B & C during months of small receipts up to October—then sell at $1–95–90–80–70-and jam prices down to D then commence again to accumulate and repeat the operation.[5]

Flint brokered the idea of a rubber cartel to the different footwear producers for several years. He gingerly pried open the jealous logjam among these longtime competitors after the New Jersey legislature enacted a liberal law of incorporation in 1889 that bypassed costly obstructions in other states and, most important, allowed corporations to hold stock in similar entities. He "avoided the ancient grudges"[6] in the industry by negotiating separately with each potential partner before forging a consolidation among the rubber manufacturers, a process the *Boot and Shoe Recorder* predicted would fail. He also employed "disinterested" capitalists from other walks of life (almost like the Knights' mixed assemblies) to provide investment funds and avoid any infighting among the principals in the particular industry. A New Haven lawyer sent a letter to the treasurer of the Candee Rubber Company informing him that Governor Harrison stopped by to agree to participate in "the proposed scheme of consolidation." The influential chief executive would leave his stock "endorsed in blank."[7] The business failure of five smaller rubber concerns in the spring of 1891 as a result of insufficient capital and underselling probably provided impetus to the scheme. Five external industrialists, bankers, and manufacturers incorporated the United States Rubber Company in New Jersey on 29 March 1892. Expert appraisers examined the value of each rubber company that formed the new enterprise and paid the participants in untested U.S. Rubber Company stock. Flint became the treasurer for almost a decade. Each firm that joined secured a seat on the Board of Directors while usually maintaining at least indirect control of the old business along with an expanded number of independent outsiders. For a time in the 1890s the Board numbered twenty-five members. The cartel suffered a rocky start and the vice president, James Ford, admitted the venture "wasn't crystallized or consolidated" for several years.[8]

Flint later told a legislative committee in New York when queried about why he purchased the Banigan properties in 1893: "Our object was to secure a business that was sufficiently large so that we could get the advantage and economies that would result from the centralization of manufacture."[9] New York State Senator Clarence Lexow caustically replied to that statement: "And incidentally to destroy competition to that extent." Flint answered simply, "No."[10] Later Flint said that he understood little about daily operations, nor could he tell the price of any of the company's shoes or boots. He also asserted that the industry had historically enjoyed "the same uniformity in regard to the price of manufactured goods."[11] On another occasion, while testifying before the United States Industrial Commission in 1902 (members included politicians and interested parties), Flint emphatically spoke very differently about monopolies. In general, he testified, they formed "owing to the war of prices."[12] One of Flint's interrogators at these publicized hearings was Charles H. Litchman, the former official of the Massachusetts Knights of Labor who had visited Millville during the 1885 strikes. He pumped Flint about his opinions on wages and unions in the rubber industry. The industrialist complained about the high cost of American labor but denied any knowledge of worker organization. Litchman must have imagined that Banigan was on the hot seat as well as Flint.[13]

The earliest members of the fledgling trust included the American Rubber Company (Cambridge, Massachusetts), Boston Rubber Company (Chelsea and Franklin, Massachusetts), Goodyear Metallic Rubber Shoe Company (Naugatuck, Connecticut), L. Candee Company (New Haven, Connecticut), Lycoming Rubber Company (Williamsport, Pennsylvania), Meyer Rubber Company (Milltown, New Jersey), New Brunswick Rubber Company (New Brunswick, New Jersey), New Jersey Rubber Shoe Company (New Brunswick, New Jersey), and the National India Rubber Company (Bristol, Rhode Island). Both Banigan and Holt met and worked at the latter firm in the 1860s when it was located in Providence and owned by the Bourn family. The current proprietor, Samuel P. Colt, would later serve as president of the U.S. Rubber Company shortly after Banigan. The nascent pool acted more as a holding company for the various facilities during its initial tenure than as a seasoned cartel with layers of specialized management. Although subsidiaries signed documents prohibiting the establishment of competing firms, the contracts failed to hold at times. All the members of the cartel agreed to purchase crude rubber supplies from the import firm, the New York Commercial Company. Flint initially controlled that raw material for profit and in the belief that a consolidated effort the size of the U.S. Rubber Company could bring stability to rubber import prices. He also cited economy of scale and centralization as the way to produce $2,650,000 in annual savings. He should have added, at least with the participation of Woonsocket; the greatest export outlet to a global market in the industry.[14]

The U.S. Rubber Company faced one very grave problem: it comprised only nine of the fifteen major rubber footwear producers. Although the would-be monopoly boasted of almost $13,000,000 in assets and annual earnings that topped $1,000,000 the year before the consolidation, the two largest and most influential hostile independents stayed out. The Boston Rubber Shoe Company, founded by the legendary E. S. "Deacon" Converse (not to be confused with the similarly named Boston Rubber Company) led the industry in sales in 1890 with $5,000,000. Banigan's Woonsocket Rubber Company, now sporting the most modern boot and shoe production facilities in the world, took second place that year with gross income of $3,500,000 but with a higher profit margin on the boot side of the industry. Banigan used more crude rubber than anyone else, mainly because of the amount that went into bootmaking, but he also sold some of the raw produce to Converse. The nine enterprises that formed the U.S. Rubber Company marketed slightly less than the Boston and Woonsocket firms together. Flint, in his autobiography, played down the inability to merge the two largest rivals immediately and added some hyperbole to the story: "This consolidation was very much larger than either of the two individual companies above mentioned and it ultimately absorbed them both."[15] Still, Flint felt obligated to give a speech on the subject in Banigan's backyard at the Providence Commercial Club on 30 April 1892. The *Boot and Shoe Recorder,* antagonistic to the combination, warned at the time that: "Brother Converse and Brother Banigan will continue supreme and independent in their respective corporations, and will do their best to make things interesting for the generalissimo who will control the destiny of the United States Rubber Co."[16] Banigan positioned himself perfectly in the years leading up to the formation of the cartel. He probably could face down the competition on his own as long as he retained control of his crude supplies. The *Boot and Shoe Recorder* labeled Banigan's rubber empire a trust unto itself. Furthermore, Banigan had augmented the sparkling facilities at Millville and Woonsocket with a new venture in early 1893, the Marvel Rubber Company.[17]

Banigan started that enterprise in Woonsocket just prior to his courtship by the U.S. Rubber Company. The plant produced the only mechanized, molded rubber shoe on the market. Banigan had dreamed of such a laborless product for years and eventually found the ingenious wherewithal in a young inventor at his plant, H. J. Doughty. He finally reached manufacturing nirvana: footwear almost untouched by human hands. The company held an exclusive patent. A series of automated valves and hydraulic actions virtually eliminated the need for production workers. The new process could mold and vulcanize a complete shoe under twenty tons of pressure in a startling five minutes. His son William B. Banigan, a Manhattan College graduate, supervised the facility. The name came from one of the company's selling agents, who remarked that the product was a marvel of beauty. The *Boot and Shoe Recorder* described it as "the handsomest rubber shoe made."[18] The Providence Chamber of Commerce, continuing to warm to its only Catholic member, also waxed ecstatic: "The little State of Rhode Island, if it continues to develop the manufacture of rubber goods, will be as important to the rubber merchants of Brazil as the tree that bears the raw material."[19] Banigan had pushed the envelope of footwear technology as far as it could go without piercing the scientific framework of production capabilities. In a way, the Marvel Rubber Company eviscerated the memory of the Knights of Labor who had sensed the coming of automation and monopoly and tried to frame an organizational structure to counter it. The union attempted to preserve its place in the distribution of profit, but the labor movement slowly retreated before the onslaught of self-operating equipment on a simple assembly line. Banigan had his revenge, although hiring actually increased because of overall expansion despite the mechanization and the ever-shrinking work time to complete footwear. The new endeavor operated out of the original Woonsocket plant dating from the 1860s. The U.S. Rubber Company now faced a competitor so advanced as to potentially serve the *coup de grace* to silence the upstart combination before it could effectively establish operations.[20]

Empire for Sale

The directors of the trust knew they had to tame Banigan, albeit on the rubber king's terms. The *India Rubber World* wrote that "there was a thorn in the side of the United States Rubber Company so long as Joseph Banigan remained on the outside."[21] In fact, Banigan cut prices upon the formation of the cartel.[22] In an interview in the *Providence Telegram* newspaper, which he owned, Banigan sarcastically remarked that "these companies have for some months been arranging to form the new trust. They are the weaker manufacturers, who have probably found it hard work to get along while the speculators in India-rubber have been juggling with the market. The manufacturers are now anxious to get out of the difficulty which the gamblers have made for them."[23] The *Boot and Shoe Recorder* pointed out that "Woonsocket is a word that one gets hold of readily and does not easily forget."[24] A cloaked appraisal conducted by Samuel Colt, legal director of the cartel, and Robert D. Evans, the second president of the enterprise for six months before Banigan's ascension in 1893, provided a glowing description of the properties. The boot factory at Millville, despite a decade of use, remained the industry's leader. The recently constructed Alice Mill shoe factory was unparalleled in the world. James D. Ford, a vice president of the U.S. Rubber Company, grilled by the same New York legislative investigation as Flint in 1897, replied to a question about the purchase of the Banigan property: "Well, the Woonsocket Rubber Company boots were sold all over the United States; the other companies would have a trade in one State; and another company in another State; but these had a universal trade; it was a desireable [*sic*] thing to own them."[25]

Flint and his Board of Directors then swallowed their pride and made Banigan an offer he could not refuse. They purchased his domain at double the face value, $9,200,000, in equal installments of common and preferred stock of the United States Rubber Company at a 3 to 1 ratio. Although Banigan was not the sole stockholder, he controlled a majority of the paper. The property included the facilities at Millville and Woonsocket and the recently organized Marvel Rubber Company, as well as the Lawrence Felting Company, which commanded $875,000 in cash. They also acquired the rubber king's devulcanizing plant to recycle rubber; nineteen double tenements in Millville; and land and water power rights on the Blackstone River. Banigan stipulated in the purchase agreement that his auditors could examine the books of the U.S. Rubber Company to determine if any other member concern had received more than Banigan, compared as a percentage of assessed value and, if so, he would be further compensated. Banigan actually discovered such a purchase and received extra payment. The U.S. Rubber Company would also name Banigan president, "to give his best efforts in the management of the business."[26] As the top officer he received an annual salary of $25,000 augmented by smaller stipends of $5,000 as general manager; $5,000 as an executive board member; and an additional $15,000 as president of the Woonsocket Rubber Company. The combined annual compensation totaled $50,000, a hefty remuneration at the time, and independent of his other investments. Banigan's son John, an executive at the Millville plant and president of the Catholic Knights of America in that town, also joined the directors, as did Walter Ballou, the selling agent for all Woonsocket goods. Banigan kept his import business intact for supplying "certain interests which he now has in rubber manufacturing outside of boots and shoes, and excepting also that he may buy rubber in the name and for the use of the American Wringer Company."[27] (He had turned the old Bailey Wringing Company into a successful monopoly although on a much smaller scale than the U.S. Rubber operation.) Samuel Colt co-signed the contract.[28]

In his own testimony before the Industrial Commission, Charles Flint disclaimed any knowledge of the price of rubber footwear. He probably told the truth. He was the ultimate outsider, providing a loose formula for monopoly and combination that would fit almost any industry. His dispassionate, evolutionary approach to cartel building possessed positive features beyond the negative connotations of Progressive Era interpretations. For example, the collective purchase of crude rubber let the air out of extreme speculation. He truly believed in the economic benefits of the model. How industrialists employed that method in regard to employees and consumers opened another chapter altogether, although he demonstrated great care and sincerity in public relations. If Flint only superficially knew the details of the retail trade in rubber, he mastered the psychological profiles of regional captains of industry. He overcame the bitter resentment among rivals, minimizing grievances that sometimes stretched back a couple of generations to the genesis of the trade. He even encountered some animosities between the fathers of the current operators.

Initially, Flint created a hybrid monopoly where owners joined a larger network but still controlled their corner of the world. His pattern evoked a painstakingly slow process whereby attrition would eliminate the individualism of each company until the parent combination ingested the parts to create a whole organism. The overall pace would be slow, but Flint accelerated prudently whenever he perceived a lack of resistance or a unified commitment by the directors. He arranged the purchase of the on-again, off-again, antiquated and lead-

erless Para Rubber Shoe Company, headquartered in South Framingham, in 1892. A year later Flint sold the property with a contract provision that the facilities could not be used for the production of rubber goods so as not to compromise the good name or good will of the
U.S. Rubber Company. Flint also bought the cut-rate, low-grade
Brookhaven Rubber Company on Long Island, New York, at a virtual
fire sale. He basically destroyed the equipment and ended operations
at the industry's cheapest producer of footwear at a loss to the cartel
of $200,000, in order to "protect the public from such a deception."[29]

If Flint failed to distinguish between the price of zephyr, lumber-
men's, or plain arctic boots, Joseph Banigan could recite, chapter and
verse, every nuance and peculiarity the footwear industry offered like
a minister quoting the Bible. Master of minutiae, Banigan predicated
his career on encyclopedic knowledge of the rubber trade. If he failed
to outline a theoretical map of monopoly, he instinctively understood
the dynamics. Banigan employed his practical understanding of rubber with unbounded, almost superhuman, energy to get the job done.
If the task involved competition, the greater the enthusiasm he displayed. Flint acted in a larger economic zone as the official cartogra-
pher for primitive monopoly; Banigan retained the raw rubber of a
lifetime under his fingernails. In theory, the partnership should have
sizzled with complementary skills and insight. Banigan, however, could
brook no pretenders to the throne. He was the rubber king. The cartel courted him; now they would have to crown him. He mastered the
global marketplace for raw material and outlets. He would bring his
former competitors to heel without an ounce of Flint's gentility. The
refugee from the Irish Potato Famine would take no prisoners in his
quest to transform the U.S. Rubber Company into a larger version of
his Woonsocket entity. Banigan had no room for Flint's patience and
evolutionary approach. The Irish-American revolutionary took office
on 12 May 1893 in the middle of that year's economic panic.

President of the United States Rubber Company

Banigan's first twelve months as president of the United States Rubber Company were as troubled as the nation's finances during that initial recessionary term in office. The chief executive's unstinting resolve
to change the nature of how the cartel managed business irritated the

partners individually and collectively despite the fact that Banigan introduced sound management procedures. In his first annual report to the Board in April 1894, the battle-weary president listed an impressive number of actions he undertook. He overturned traditional, static ways when he reduced the carrying charge of 2¾ percent to 2 percent for crude rubber to Flint's New York Commercial Company, "thus saving $35,344 per year for our Company." Incredibly, Banigan had held major stock in Flint's enterprise but must have felt that jeopardizing his holdings was warranted although the paper was hidden under the name of his son-in-law. Furthermore, he dragged Samuel Colt along to audit the importer's books and uncovered "errors which had crept into their methods of buying rubber" and, "after long and trying efforts," saved the combination another $25,000 annually. Banigan, in a further shakeup, imported crude rubber at a lower price for the trust using his own credit. (In a subsequent trial Banigan testified in court that Flint kept two sets of books.) Undaunted, despite the humiliation he bestowed on Flint, who had made and broken governments in South America, Banigan announced that, "since this settlement the purchase of rubber has gone on in a more pleasant manner, but more than half of the rubber consumed by your Companies has been imported by me with my Credits and at my risk, upon which they receive their full commission." Banigan seized the upper hand morally and profitably while placing the trust's founders in a defensive posture. Flint resigned from his position with the New York rubber importer but not from the cartel.[30]

Banigan's frontal attack on Flint and his desultory progress and questionable methods occupied only the first paragraph of the seven, single-spaced typewritten pages of his first annual report. In order to further integrate the still disparate enterprises, he reported, Banigan formed the United States Rubber Advance Club. The organization consisted of the various supervisors of the member firms who traveled with Banigan to inspect each facility's "machinery, buildings, tools, and appliances." During periodic meetings, the supervisors implemented efficiency measures gleaned from the different tours. Banigan carped that "the labor-saving appliances possessed by any one Company have been adopted by each of the others, and the savings already effected in this direction will amount to many thousands of dollars each year." No doubt the club members must have taken very careful notes at the Woonsocket enterprise, their first review. Later, in a court statement, Banigan said that, "when we attempted to stop or change certain small mills that were not earning money, we were met with the statement that they had an agreement with Mr. Charles R. Flint that he should keep that mill running all the time—that it wouldn't be stopped." He also claimed that executive board members feared the supervisors would usurp the board's prerogatives.[31] Not included in the report was a similar meeting that Banigan hosted in Woonsocket for all sales agents. He made them take a course on ethics so as to provide realistic product appraisals to wholesalers, a move applauded by the *Boot and Shoe Reporter.* Even vice president Evans visited. Banigan also instituted a uniform system of packing footwear. Flint must have cringed at this micromanagement of the giant operation.[32]

The examination of Colt's former National India Rubber Company in Bristol added some humor to the usually tense atmosphere of these expeditions. Banigan inspected "an old press in the molding department," much to the worry and chagrin of local management, who knew of his obsession with modern equipment. "Gentlemen," Banigan remarked, "there is the beginning of my experience in the rubber business. On that press I worked making rubber curry-combs when a young man." The machine had been hauled along when the company moved from Providence to Bristol under the administration of the Bourn family before Colt purchased the business. Banigan told of paltry wages and his struggle to learn the trade from those humble beginnings. While operating the mechanism at the Providence plant he met John Holt and his daughter, Margaret, his future wife. Flint possessed a pedigree in finance but not one in the genealogy of rubber.[33]

Banigan's initial report to the board, and eventually to the stockholders, was an outline in progress. He continued to tell of improvements. He eliminated the use of wooden lasts and boot trees, shaped like footwear, which molded and formed the products. These ubiquitous instruments shrank and crumbled easily, resulting in uneven size and costly replacements. The superintendents suggested the substitution of sturdier sheet metal versions that actually cost less and produced uniformity. Banigan also spoke of the high cost of cotton liner, which in its current form resulted in great waste as well as a variety of lining styles—111 variations at Millville alone. He established a small textile operation at the original Woonsocket plant. He found the savings so impressive that he advocated a separate facility to produce from scratch the cloth to supply the felting firm in Millville, eliminating one more middleman—always the hallmark of his own operations. He also bought a controlling interest in the Hammond Buckle Company in Rockville, Connecticut, to supply adornments to footwear, "thereby getting the very best article on a confirmed standard of quality at its actual cost." Furthermore, in line with some of the above changes, he implemented general standards for a variety of styles while eliminating certain fashions and overlapping production of closely related goods. The one area that he tackled gingerly involved economy of scale at facilities of various sizes: "I, therefore, strongly recommend that the output of the smaller Companies be made at the larger Companies, thereby increasing the product of the latter, thus reducing the cost of production." Although he would shift minor product lines to the antiquated factories, such suggestions sent shivers among some of his old competitors, who saw the handwriting on the wall, and had apparently consummated earlier agreements with Flint to keep their concerns running. Flint probably agreed with the direction of these changes but not the pace; he had wanted such closures when the veteran owners retired or died.[34]

Banigan, not totally insensitive to the feelings of others, endorsed Flint's earlier actions to close down an underperforming facility and a footwear producer of inferior grades, although in tough economic times the working class could afford only the cheaper variety. He addressed the financial conditions in 1893: "While the general depression of trade throughout the country has naturally affected all business, it has not affected our business quite as badly as it has others." Historically, the rubber industry had a long track record of financial undulations governed by the weather that steeled it to economic downturns more than most commercial establishments. On the heels of the recent financial panic, Banigan instituted a price reduction in the spring to take advantage of cheaper shipping rates in the summer when wholesale jobbers took footwear orders for retail sales. He ended his detailed and very demanding report with this concise statement of the essence of his presidency: "There are too many factories for the production of rubber boots and shoes, and this will be particularly true when the improved methods are fully adopted." Although the Board disingenuously approved the controversial report, they balked at the inclusion of one rather personal item. Banigan reported that, pursuant to the purchase agreement for the Woonsocket facilities, he had the right to inspect the cartel's books to determine if any other member received a greater payout percentage. Robert D. Evans, the second president of U.S. Rubber, just before Banigan, apparently tried to increase his share of the pie without the requisite contractual language that Banigan included in his own agreement to sell. Banigan squelched his predecessor's undocumented deal. The Board stuck that part of the report in an appendix and kept it from stockholders. Banigan had stirred a hornet's nest of grumbling, including the alienation of two of the top officers, Flint and Evans.[35]

At the time of Banigan's first presidential report, the U.S. Rubber Company had balanced assets and liabilities that both came to just over $40,000,000. In the Panic of 1893 the cartel had a small surplus of about $250,000 but, more important, it paid the 8 percent interest dividend on almost $20,000,000 in preferred stock. In his second annual presentation in May 1895, Banigan reported a better financial climate despite the lingering downturn. The surplus swelled to almost $1,000,000 on about the same amount of money on the balance sheet as a year earlier. The company also paid a 2¼ percent dividend on common stock as well as the 8 percent on preferred. Despite the fair weather in the West and Northwest, "in the Eastern and Middle States we were favored with plenty of snow and rain, so that on the whole, we may consider that we have had a little better than an average winter." He also pitched the layout and improvements of the newer facilities, which resulted in a reduced insurance premium of 30 cents per 100 dollars of evaluation instead of 50 cents.[36]

In his first annual statement in 1894, Banigan started with his activist agenda as it related to savings before actually examining the bottom line. In the second installment in 1895, he discussed the encouraging finances of the enterprise first. The rest of the report reflected his deep disappointment in the cartel's inertia: "Of the many improvements recommended in my last Annual Report, very few have been adopted." Banigan had no experience with subordinates who failed to follow orders. He carped that business had continued as usual, with little regard to economies of scale and consolidation of production. The other directors still kept watch over the companies they had once nurtured and had no intention of disassembling their related facilities. Banigan scolded them: "In other lines of business such methods have been pursued, small and out of the way factories have been closed and their product made at the larger and more improved ones, and it has always worked to advantage." Nor was Banigan shy about which branch of U.S. Rubber should increase output: "A savings of over a million dollars a year can be made by running the two large mills of the Woonsocket Rubber Company to their full capacity, and three hundred days a year." Miffed at the underutilization of his modern facilities, Banigan's assessment of lost potential income showed up in the appendix, again shielded from stockholders.[37]

Boldly outlining his vision in greater detail, Banigan proposed that the largest divisions of the company—the Woonsocket, Goodyear, American, and Candee plants—undertake primary footwear production while either the American or Colt's National facilities manufacture the cloth and miscellaneous goods. The first four already produced 80 percent of the monopoly's footwear. He advocated closing the other component parts of U.S. Rubber. "While there may be some sentiment in favor of running each mill as formerly, yet it should not be permitted to interfere with the success and prosperity of our Company." Banigan then threw out the bone: "To consolidate the production of the four Companies above mentioned, and the classification of production to some extent, will add a million and a half dollars per year to our net earnings." He also outlined plans to have the parent company control more of the selling function of the others, reducing the duplication of costs in that area as well. Flint must have winced to watch his own agenda being implemented in machine-gun fashion instead of in his own breechloading style. But the Board, unbeknownst to Banigan and in his absence, had held an irregular, secret meeting before the second annual presentation in order to increase the number of Executive Board members and to consolidate their power to counter the president by adding sympathetic colleagues. In the lawsuit that followed, Banigan's counsel labeled the gathering "a movement against Mr. Banigan."[38]

Resignation

There would be no third annual report by Banigan. He quit just before the end of his three-year tenure, in 1896, although Woonsocket sources had predicted the outcome more than a year earlier. Banigan received only a partial salary the last twelve months, a reflection of the board's disenchantment with the president, and a sign of the subsequent drift of the company. Apparently the board met clandestinely again at the Waldorf Hotel in New York City and decided to force him out. His disgruntled successor, who was also his predecessor, Robert D. Evans, ironically complained about the failure to consolidate production in his own truncated report in 1896. He admitted to the board of directors that no facility was fully utilized and outlined plans that differed little from Banigan's vision. Evans still reported a $400,000 surplus.[39]

With Banigan's announced withdrawal, rumors swirled that the syndicate would now close Samuel Colt's Bristol plant altogether. Apparently some directors wanted to give Banigan a taste of his own medicine, regardless of the viability of their plans. After his departure they foolishly targeted both the Woonsocket and Millville plants for termination. One report claimed the new board at U.S. Rubber discovered that "Banigan's factories were out of date," a ridiculous assessment to anyone not caught up in the verbal battles being waged among board members.

In a last gasp attempt to salvage the situation, Banigan had put together a "harmony banquet" in New York City, but to no avail,[40] and he tendered his resignation at the banquet. He could not control the mutiny despite the soundness of his projects. Still he was elected a director, but he refused, declaring: "I do not desire to take any active part in the management of the company; but I am willing to give any advice if asked, and will render all assistance in my power to the betterment of the company, as I represent the interest of widows and orphans, and from personal knowledge and contact with many other lines of business I believe the rubber business to be the best paying industry in the world to-day."[41] The reference to "widows and orphans" undoubtedly referred to stock used as an endowment for his charitable creations in Rhode Island. Despite his withdrawal from the U.S. Rubber Company, Banigan would not retreat from the rubber industry. He would take on his greatest challenge: direct competition with the retrenching trust in a new venture that employed his considerable fortune and skills. Banigan purchased a former textile factory in Providence, in the industrial section of Olneyville, during a time of high unemployment.

Knowledgeable observers still felt that he might attempt some sort of back door takeover of U.S. Rubber. Common stock in the syndicate fell from 60⅜ in 1893 to 16½; preferred went from 99 to 74 as the result of the turmoil surrounding his departure. Ironically, the trust began the painful shutdown of underperforming facilities that Banigan had advocated, including, astonishingly, the Marvel Rubber Company in Woonsocket. As a parting shot, Banigan sued the trust for salary withheld during his final term, an action that would create an interesting drama by itself.[42]

The Trial

The trial for Banigan's back pay had very little to do with the money; his annual salary from the syndicate had been reduced from $25,000 to $10,500 (although the other stipends totaling $25,000 remained in effect), for a net loss of $14,500. Both sides wanted to skewer the other in preparation for a public relations campaign that would ensue when Banigan commenced production. The trust desperately needed to avoid the stigma of being an insensitive moneymaker and, at the same time, prevent him from taking the high road to gain public sympathy as the executive who stood up to corporate shenanigans "in the interest of widows and orphans." Because Banigan sued to recover lost salary, his lawyers attached the property of the syndicate and crossed swords with his fellow board members in the Rhode Island Supreme Court. He engaged the prestigious Yankee law firm of William R. Tillinghast in Providence, probably unable to find a local Irish-American law practice with experience in the high altitude combat of corporate America. The U.S. Rubber Company hired the legal mastermind of the Republican Party machine in Rhode Island, William G. Roelker. Although both parties battled over nuances and perceived loopholes in the contractual arrangements, the real issue focused on the morality of the emerging Progressive Era. The opposing lawyers each portrayed the other's client as Mammon. Despite the rancor, both sides must have made a silent agreement to bypass certain events during Banigan's presidency that would stain plaintiff and defendant alike. The case began in January 1897 in Providence at the Common Pleas division of the state Supreme Court, without a jury.[43]

Roelker started the proceedings by explaining why the cartel paid the combined $50,000 to Banigan in the first place, "although it was believed to be a much higher salary than had ever been paid to anybody in this country in that business before."[44] The board of directors anticipated great things from Banigan's leadership after his "assurances" to them about his capabilities. Two years later, according to counsel, "the expectations which he had held out to them had not been realized"[45] and the board trimmed his wages. He then accused Banigan of secretly unloading U.S. Rubber stock in the guise of loans and thereby securing several million dollars and undercutting his loyalty to the firm by spending that money and company time making other investments, including the preparations for a rival to the U.S. Rubber Company. After his opening statement, Roelker called a cavalcade of rubber moguls to the stand.[46]

Robert D. Evans testified that the company had been extremely anxious to secure Banigan's services. The cartel virtually bribed him with perks, including the $15,000 annual stipend from the Woonsocket Rubber Company—another indication of how loosely the combination had been drawn by Flint, allowing former executives still to draw compensation from their own enterprises. Evans said that Banigan had promised to escalate the price of preferred stock to $125 a share and not sell any of his own holdings until reaching that figure. Banigan, according to the witness, allegedly lost interest after the first year although he signed on for three terms. Tillinghast later recalled Evans and asked him about Banigan's first several months in office. Evans claimed that Banigan went to Europe for a few months on "holiday" during that period.[47]

In reality, during this period of financial panic in the United States, Banigan had met with another emissary of U.S. Rubber, a close associate of Flint, in order to secure loans from British capitalists. Evans denied any knowledge of the arrangement. Tillinghast read coded cables to Banigan from U.S. Rubber about the dire necessity of obtaining sterling to meet maturing paper. Banigan also provided a 100,000 pound note in the coin of the global realm in that era from a lending house in London and also arranged for a shipment of African rubber. Later in the case, when Banigan testified, he divulged that he lent the Woonsocket Rubber Company and U.S. Rubber some $1,500,000 during the Panic. He claimed that $100,000 remained outstanding, which led to another court case that was eventually dropped.[48]

Henry L. Hotchkiss, president of the L. Candee and Company Rubber Company in New Haven, testified next. He had spoken against Banigan in the rubber shortage trial in 1887 involving the import house of Shipton-Green and had served with Flint on the Board of the New York Commercial Company that Banigan clipped in his initial term. Hotchkiss, an original director with the rubber syndicate, reiterated basically what Evans said.[49] Hotchkiss, who was also chairman of the committee on accounts, did recount his suspicions that Banigan received unreported but eventually unsubstantiated "commissions" from the Woonsocket Rubber Company. Banigan's lawyer hammered away at Hotchkiss about such unfounded suspicions and rumors "in the air," which weakened the credibility of the witness.[50]

James V. Ford followed the tentative Hotchkiss to the witness stand. He ran the Meyer Rubber Company in Milltown, New Jersey, and served as vice president of U.S. Rubber. He reiterated Banigan's pledge to increase stock values, thought his overall compensation excessive, and believed his involvement waned during his last year in office. Tillinghast, Banigan's counsel, got Ford to admit he served several corporate masters himself in addition to the rubber syndicate, the same charge the defense was leveling against Banigan. Ford belonged to the board of directors of four other companies, all requiring attendance at frequent meetings. When asked the particulars of Banigan's inattention to duty, Ford answered, "I don't recall any thing just now."[51]

The U.S. Rubber Company lawyer tried valiantly to prove that Banigan made formal arrangements for other personal enterprises before actually stepping down as president. Roelker questioned the owner of a textile mill in Providence that Banigan purchased, to determine the actual date of the sale. He also grilled the treasurer of the Woonsocket Rubber Company to see if Banigan received compensation after his resignation. Banigan proved too wily to be snared by authenticating documents for another business venture before his term expired at the cartel. In another blow to the monopoly's arguments, Colt described several secret meetings of the executive board to undermine Banigan's power and authority.[52]

At a continuation of the hearing on 24 February 1898 Roelker called Samuel P. Colt to the stand. Colt affirmed Banigan's $50,000 in total salary but considered it "a very liberal compensation."[53] As earlier syndicate witnesses had done, he explained that he had been influenced by Banigan's promise to lift the preferred stock to $125 a share and his pledge not to sell any of his own shares beneath that ceiling. He also stated that Banigan's participation fell off after the first year in office, especially during the third term, as the cartel's legal team reinforced the perception that Banigan may have been too involved in other outside projects by that time and sold much of his stock through intermediaries. As a Rhode Islander, Colt spoke more specifically about Banigan's other interests, especially office buildings in Providence and investments in Utah. Colt, who originally inspected the firm's books with Banigan during his first term, admitted the enterprise kept two sets of accounts, although he tried to avoid the issue of impropriety. But he was caught between a rock and a hard place. Tillinghast queried Colt directly:

> Question: "Now I ask you as a director, and as the legal director and a member of the Board of Directors and executive committee, whether you understood then that Mr. Banigan was under obligation to hold his stock for any length of time?"
> Answer: "I don't suppose there was any contract about it."[54]

Tillinghast had cornered Colt. He further wounded him, accusing the politico of trying to buy Banigan's stock himself. Although Colt demurred, the testimony seemed half-hearted and probably destroyed the trust's argument that Banigan promised not to dispose of his shares. In fact, Colt admitted that other officials had sold smaller amounts of stock. Colt, Rhode Island's former attorney general (1882–1886), had been humiliated on the stand by his inconsistencies, memory lapses, and inability or unwillingness to provide much credible detail. He later ran unsuccessfully for governor and senator while president of U.S. Rubber.[55]

Until he spoke for himself, Banigan had no champions testifying on his behalf. He was the only Catholic or Irish witness. The management team of the United States Rubber Company testified in hostile concert. The Roelker defense team reiterated several salient arguments to convince the old-stock judges of the Rhode Island Supreme Court that Banigan deserved less salary during his last year at the helm of the cartel. U.S. Rubber had paid him a princely sum to manage the company because of his perceived ability. Banigan whetted their appetites further when he promised to enrich the stock and not sell his own shares below that figure. When Banigan failed to deliver and actually dumped his holdings rather surreptitiously, in their eyes, he lost interest in the syndicate and turned his attention to other personal projects including the formation of a rival rubber company.

However, the collective defense failed to muster the required credibility or unity of agreement among the directors. Contradictions in testimony hurt the monopoly's case. As to the reduction of presidential salary during Banigan's last year in office, the reasons appeared vindictive more than strategic. Permeating the trial was a sense of the larger issue of the upcoming battle between the two forces although that engagement remained, in a phrase uttered by Banigan's lawyer for something else, "in the air."

Banigan was the final witness. He unequivocally demonstrated that the cartel had courted his Woonsocket empire and, with Colt, examined the facilities. "I told him they would have to pay a very high price for it; it was a good earning plant; my baby; I didn't like to sell it; as I established the company."[56] Colt and Robert Evans had met with Banigan on a half-dozen occasions in Providence at the beginning of 1893. He flatly refused to market the property or become president unless the method of importing crude rubber through Flint's New York Commercial Club changed. Flint attended a meeting and agreed to increase the stock of the Club and sell Banigan $500,000 of the importer's shares; Evans, $200,000; and Colt; $100,000. Those numbers, combined with the cartel's investment of $400,000, provided the syndicate with a controlling interest in the New York Commercial Club in more ways than one. Banigan achieved an agreement on the spot to lower the Club's commission on imported rubber from over 2 percent to just 2 percent. When negotiations stalled over the price for his properties, Banigan walked out of the meeting and left the U.S. Rubber team alone one night. Evans went to Banigan's home just before midnight, accompanied by Banigan's son John, to accept the rubber king's terms. The parties signed the purchase on 22 March 1893 after several auditors swarmed the Woonsocket ledgers.[57]

Banigan's testimony, from the start, included greater detail and recall of events than the mostly terse replies of previous syndicate witnesses. The rubber master, with no formal legal or educational training, seemed to forget nothing when it came to discussing "my baby" or the rubber industry in general. His life in the business became the family scrapbook, easily recalled and explained. He also disarmed the judges, perhaps, with his honesty, admitting that he asked for an above average salary to replace his loss of the annual $60,000 he earned importing crude rubber. Banigan denied ever promising that he would grow the stock to a certain threshold or refuse to sell his own investments below that number. Because the cartel's witnesses hesitated and spoke vaguely about what he had or had not promised, Supreme Court justice Charles J. Matteson interjected to declare: "I don't think that they claim that there was any agreement or any promise ever stated on his part," a remark that must have made Roelker wince in legal pain at the loss of a key argument. Banigan went on to say that he may have "used the $125 for the preferred stock or not. . . . But it was always with the understanding that all should work together; and I believe it just as firmly to-day as I believed it then—that if they had worked together the stock, common and preferred, would be sold at par or above."[58]

The Joseph Banigan Rubber Company

Banigan moved at lightning speed to take advantage of the unstable situation in the rubber industry just after his own lawsuit and before Charles Flint had to testify before the New York legislature's investigative committee. He incorporated the Providence-based Joseph Banigan Rubber Company in December 1896, mere months after his resignation, which coincided with the end of his legal obligations not to compete with the cartel or steal any former customers of the rubber trust. The new enterprise occupied a 3.6 acre site at the corner of Valley and Eagle Streets in the former Saxon Worsted Company. He exploited U.S. Rubber's new practice of not paying salaries to supervisors and other officials at the Woonsocket and Millville plants while those enterprises remained closed. Banigan swept in and reassembled his veteran management force. Walter Ballou, Woonsocket's sales agent, who had joined the boss on the trust's board of directors, left with his longtime colleague. Samuel Colt testified later that Banigan lured away "a number of the foremen, and more or less of the help."[59] Moreover, he had always catered to the crucial independent wholesale jobbers who peddled the footwear from factory to retail outlets by carrying their debts for a reasonable time period. That old alliance paid

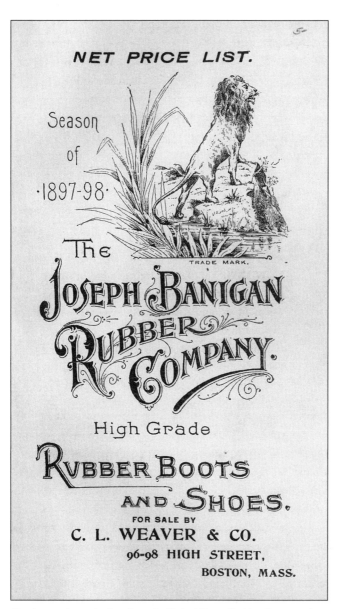

Banigan, after resigning from the United States Rubber Company in 1896, used his own popular name in the footwear industry for his independent venture. (Author's collection)

dividends once again. Furthermore, Irish machinists, bootmakers, and other unemployed rubber hands joined the exodus from Woonsocket and Millville to regroup with their original boss regardless of past grievances, especially in light of the impersonal and corporate behavior of the cartel. Operatives may have nurtured a greater appreciation of working for their own kind, an Irish compatriot, instead of the new Yankee owners. Published reports highlighted Banigan's behind the scenes attempts to help his veteran workers during his tenure as president of the U.S. Rubber Company. He "naturally desired to protect his old employees."[60] At the same time, many of his fellow cartel directors tried to accomplish the same thing by sheltering their former facilities and workforce at mills that Banigan considered redundant and expendable.[61]

Déjà Vu

Banigan's failure to warn his operatives of an impending wage reduction had triggered the original strikes in 1885, although the walkout reflected many other problems as well. Then, the young and militant Irish-American operatives, perhaps prematurely and unwisely, left work because of Banigan's perceived disrespect. Eleven years later, as Banigan struggled with his latest competitors in the U.S. Rubber Company boardroom, those corporate adversaries ordered a temporary shutdown of the facilities at the Woonsocket Rubber Company behind Banigan's back as he was losing control of the trust. The still predominantly Irish employees in the Millville plant arrived at work on the morning of 15 January 1896 only to discover a few hours later that the mill would close at noon and operate only intermittently to fill special orders for the foreseeable future. Some of the same skilled cutters and bootmakers who had jumped the gun in 1885 and left their work benches in protest, now meekly bowed their heads and inquired when the furlough might end. The lack of any forewarning in 1885 and 1896 initiated very different responses as the temper of the times changed dramatically in a decade. Caught between a recessionary storm, battling rubber magnates, and an impersonal employer, workers counted their blessings and hoped the layoff would be brief. The operatives, once unified against a systematized Irish-Catholic family business, were now weakened by the muscle of corporate management. Father Michael McCabe and John Holt, the toilers' erstwhile allies in 1885, were no longer in the picture. The Knights of Labor, which made history just a decade earlier during the Great Upheaval, had already passed into history. Some of the 1,000 skilled factory workers tossed into unemployment at Millville in 1896 had sought an antidote to corporate one-sidedness for years but came up short in the union political intrigue that infected the Knights in the late 1880s, a scenario similar to the conspiracies in the U.S. Rubber Company in the 1890s. Ironically, these workers now looked to Banigan for help. Like the poor peasants and farmers in County Monaghan desperately petitioning the landlord Shirley and his agent Trench during the Famine, rubber operatives now distastefully sought protection in the paternal cradle.[62]

The Woonsocket newspapers announced each new departure to Providence: "the removal of the rubber help from the village [Millville] have averaged over a dozen a day for the past few days and perhaps averaged a third of that for a number of weeks."[63] Banigan soon advertised that "the most experienced rubber workers known in the manufacture of rubber boots and shoes" fashioned his products.[64] By January 1898, these displaced operatives renewed the practice of giving Christmas presents to favored supervisors and management personnel as they had years earlier. Once Banigan had renovated the old textile building on Valley Street into a state-of-the-art footwear plant (although not on the same scale as the Alice Mill), he traveled to Europe to seek new customers. Elected officials in Millville and Woonsocket agonized over the retreat. The overseer of the poor in Woonsocket reported another dismal season in the aftermath of the lingering Panic of 1893. By the spring of 1897, Millville had 175 empty tenements, unheard of in the halcyon times of a few years earlier. Woonsocket contained even more deserted dwellings. One unemployed rubber operative committed suicide there. Unable to obtain work at Banigan's new Providence headquarters, hundreds of others, according to the *Woonsocket Patriot*, actually returned to Ireland to wait out the depression.[65]

The United States Rubber Company had plenty of resources to battle its former president and prepared to flood the market with inexpensive and discounted goods to undercut the upstart Banigan, just as the rubber king himself had done when the combination started operations without him in 1892. The Boston Rubber Shoe Company, the largest independent outside the cartel fold, cut prices first and forced the syndicate to do the same, but at a reduction of dividends to 2 percent. In some ways, Banigan engaged himself in this conflict. The cartel employed his former, modernized plants, economy of scale, and some of his own tactics against him. Banigan remained the loose cannon. He was never at a loss for fresh ideas in evolving situations. Even before the consolidation, rubber companies uniformly released their price lists for the year during the last week of March. Banigan broke precedent and beat the deadline by several weeks even though actual production began only a few months earlier. If the trust envisioned a race to the bottom in price and quality, Banigan squelched that with a list of goods that cost about 6 percent more than U.S. Rubber's comparable footwear. He probably provided larger discounts to the friendly wholesale jobbers who distributed the products to retail outlets. The combination followed suit and discounted for early orders. The *Boot and Shoe Recorder* described the scene: "The campaign for 1897 opens with the Banigan battalion wheeling into line and firing a salute in the shape of a new price list." The cost of a comparable pair of shoes dropped to the level of 1892, a boon for consumers. Robert Evans, president of U.S. Rubber, admitted that the simple announcement of Banigan's incorporation created competition, months before actually creating any product. He also claimed that the new enterprise did not interfere with the syndicate's customers.[66]

Banigan forayed, for the first time in his career, into the murky area of selling second-grade footwear, but under cover of the Woonasquatucket Rubber Company, named for the river that passed by the Providence factory. He camouflaged his strategy behind his longstanding reputation for quality boots and shoes and his current position as David taking on Goliath. Furthermore, he had an independent ally. E. S. "Deacon" Converse had kept his vibrant, high-grossing Boston Rubber Shoe Company unattached from the cartel, until several months after Banigan's death in 1898. Banigan and Converse were the only manufacturers who maintained their own crude rubber supply autonomously from Flint, and rumors floated of a possible merger between the two independent companies. Several cartel directors also resigned to form smaller operations, while other independents continued to test the water, usually with very narrow product lines and a slogan such as "Not Made by a Trust."[67] Colt allegedly feared "the competition of the accumulating independent rubber factories."[68] The *Woonsocket Patriot* predicted "a merry war."[69]

In May 1897, Robert Evans said in the trust's annual report that business "has suffered to some extent from the disturbed conditions that have prevailed in the political, mercantile and financial worlds."[70] He could just as easily have been describing the dissension, boarding on anarchy, over the direction of the combination. He promptly resigned. The *Boot and Shoe Recorder* quipped about the discord on the cartel board: "Mr. Banigan, while president, met with much the same trouble and preferred retirement." The journal also reported that "each concern was conducting its business on a semi-independent basis," reversing the original centripetal intent of the combination only five years after its incorporation.[71] That was Flint's calculated gamble, the price of slowly gluing the disparate parts. About the same time, Flint seemingly joined the dispersal movement himself and opened his own rubber import company in West Virginia. And then good times returned for the footwear producers. The gold rush in the Alaskan Klondike attracted thousands of unemployed Americans, including many New Englanders. In 1898 the Spanish American War scooped up some volunteers and financed military suppliers. Both events spurred production of rubber goods and footwear, further brightening the employment picture. At the same time, Banigan still presided over his own cartel, the American Wringer Company, fashioned out of the original Bailey enterprise. In 1897 it earned almost $200,000. With times so flush, both sides in the battle for rubber supremacy could survive and prosper, although the loss of skilled workers in Millville to the Banigan plant in Providence spurred the employment of "green hands" in the Massachusetts boot operation. The trust's stock hit ninety by the middle of 1898.[72]

The revival of the boot and shoe trade fueled rumors that Samuel Colt would soon become president of the trust. Colt and Banigan had remained mysteriously close and kept communications open. Reports speculated that the syndicate might restore its satellites to their original orbit, although Banigan denied any desire to rejoin. Inevitably, his salary suit against the U.S. Rubber Company had engendered a countersuit by the trust for breach of contract, among other things. The Appellate Division of the Rhode Island Supreme Court eventually ruled in Banigan's favor:

[W]e can see no sufficient reason for the sweeping reduction of salary in his case. It is evident, however, from the evidence that a strong feeling of hostility against the plaintiff and a strong opposition to his policy in the management of the affairs of the corporation had sprung up among his associates, and we cannot resist the feeling that the action of the executive committee and of the directors in fixing the plaintiff's salary was influenced, consciously or unconsciously, by these conditions.

The justices awarded Banigan the amount of his salary reduction, and a moral victory. More important, the trial record involved testimony from many of the key players in the corporation as well as Banigan's impressive declarations from the witness chair. Like most judicial proceedings at this level, the interrogations of both sides uncovered the scintillating dynamics of the inner circle of a major corporation, replete with grand designs, personal intrigue, and downright pettiness. History offers few other primary sources so incisive and revealing. There was no consolation for Joseph Banigan, however; he had already died by the time the court rendered its decision.[73]

Banigan's case against the United States Rubber Company thus uncovered the scheming behind monopoly formation in the America of the Gay Nineties. The rubber cartel seemed petty and even criminal as testimony unveiled questionable plots to corner windfall profits, but Banigan's soiled fingerprints were found on a number of the projects as well. His financial acumen usually added a stiff carrying charge to anything he touched. On the other hand, he planned a rational strategy to dominate the footwear market and strain out the duplication in the industry. His critics told the truth when they claimed he lost interest in the syndicate after a couple of years of banging his head against the boardroom wall because his narrowly focused colleagues tried to hang onto a lost world of family business. In some ways, these still localized entrepreneurs protected the interests of their hometowns and workforces in the same manner that Banigan sheltered lifelong concerns in Woonsocket and Millville. But the Darwinian rubber king had bested them all with his ultramodern factories and techniques and felt that the spoils should go to the winner. Furthermore, he had molded the common clay of Irish America, ethno-religious outcasts rejected by most Yankee industrialists, into skilled rubber operatives. Like the

Knights of Labor, the board members of the United States Rubber Company refused to acquiesce to Banigan's business and labor practices regardless of whether he held the wild cards. The bitter contest failed to take center stage nationally because larger cartels and combinations grabbed the headlines in even greater struggles: transportation, steel, and oil, to name a few. Still, in the common mind, which was generally devoid of detailed, insider understanding of life under monopoly capitalism, Banigan appeared as the people's rebel and Americans liked popular defiance, misplaced or not. "The clear head, the admirably lucid mind, the calm, self-confidence, the unruffled serenity he displayed as a witness there," gushed the Catholic diocesan newspaper, "gave the curious public an indication of the elements that made a successful business man."[74]

During the trial, the bumbling protagonists for the United States Rubber Company—minor captains of industry—occasionally got it right, like a broken clock giving the right time of day twice every twenty-four hours. Banigan was out of the office that last term, they testified, tracking down investment opportunities in his beloved Providence or in the tangled religious web of the Mormons in Utah. They correctly surmised that he was also plotting, to initiate a serious rivalry to the rubber monopoly. Banigan, the dangerous revolutionary, failed to respect the gentlemanly past practices of the established Yankee ascendancy. He could never quite fill his belly or satiate the survival instincts engendered by the "Great Hunger" in Ireland. He never seemed to look back, but also he never appeared to be able to escape the psychic trauma of forced emigration from his homeland. The primitive and basic will to live appears to have supercharged Banigan's philosophy of financial accumulation. The phantom Famine seems to have lurked behind every deal that might go bad, in the guise of middlemen and suppliers who might default and competitors who lived by the same credo of survival of the fittest. Banigan could not rest, endorse complacency, or cease his toils.

CONCLUSION: A MEMORIAL FOREVER

· ·

> *"It was generally believed that in his business there was additional zest to the competition he encountered because he was an Irish man and a Catholic."*
> —*Obituary in the* Providence Visitor, *" July 1898*

> *"Eloquent reference was made to his private charities known only to Him who knows all things."*
> —*Bishop Matthew Harkins, eulogy, July 1898*

Although Banigan stepped down officially from his position as president of the United States Rubber Company in the spring of 1896, he had spiritually abandoned the position a year earlier as charged. He carefully avoided formal commitments to any other ventures in order to legally protect himself from any accusations by his former colleagues. They nonetheless assailed him for being more interested in newer enterprises and other financial opportunities, while maintaining his ongoing investments, such as the American Wringer Company. Banigan soldiered on, seemingly invincible. On rare occasions during his career a newspaper report might mention a trip for rest and relaxation, but the rubber king could never enjoy a vacation without some combination of business activity. The cartel characterized his fundraising visit to England during the Panic of 1893 as a "holiday" so as not to alarm investors and the stock market. Banigan seemed to avoid those psychic afflictions that temporarily felled some of the other high and mighty during the Gilded Age. Very infrequently, a mention of exhaustion or illness showed up in the Woonsocket newspapers rather than daily journals in Providence, Boston, or New York. With so many connections to his empire's home base, few secrets remained uncovered locally. Several months after he severed relations with the cartel, and as he was firing up the new footwear facility in Providence, the *Boot and Shoe Recorder* announced without any fanfare that Banigan had suffered a "billious attack."[1] How long he endured the liver and gall bladder cancer is unknown, but his indomitable will carried him on to the end without letup.

Real Estate

Although Banigan's illness would soon bring the rubber king down, he invested with the Mormons and formed the Providence Building Company in 1896. Two years later he finished the capital city's initial skyscraper, a ten-story office building that, like his factories, featured all the latest amenities: steel, elevators, and fireproofing. The structure, located at 10 Weybosset Street within the boundaries of his old Irish fifth Ward, has undergone several name changes but remains a valuable piece of downtown real estate and still features the owner's quality craftsmanship and material, which have stood the test of time. An Irish immigrant from County Monaghan sent a bird's-eye postcard of the city back home in 1901. He noted on the margin the placement of the Banigan building as if the reader in the county where Banigan was born would recognize the name. About the same time he finished the skyscraper, Banigan completed the seven-story Alice Building at 236 Westminster Street, a structure named after his daughter. The property featured a warren of ground-level shops. (Currently listed on the National Register of Historic Buildings, the edifice underwent an $8,000,000 renovation into urban apartments in 2002.) Both buildings towered above the site of his boyhood home on Orange Street. In the same vicinity he also purchased the Wheaton and Anthony (office) Building, built in 1872 at 65 Westminster Street. Elsewhere, he owned the valuable, block-long Austell Building in Atlanta, Georgia, which he acquired after a loan default and a sheriff's sale only a couple of months before his death. He and his wife traveled there on a six-week excursion in the spring of 1897. He may have recognized that lucrative venture while visiting his sister who resided in the city. Earlier he purchased property after the Great Chicago fire and made a small fortune.[2]

Most of all, during his few remaining years, Banigan injected energy and capital into his latest rubber entity to battle the cartel. By the

Built in 1896, the Banigan Building in downtown Providence was the city's first skyscraper. It remains a prime piece of real estate after a recent renovation. (From the Manual of the Rhode Island Business Men's Association, *Providence, 1907)*

beginning of 1897, the Providence facility, with a workforce of 750, produced almost 8,000 pairs of shoes and an additional 2,000 to 4,000 boots. The output drew accolades and orders. Also in early 1897, the Boston Rubber Shoe Company announced price reductions that seemed to presage the centrifugal self-destruction of the rubber monopoly. Banigan's archenemy and successor at the syndicate, Robert D. Evans, resigned a mere six months after stepping in to replace him. Ironically, Evans criticized the dissension among board members that he had helped create. The *Boot and Shoe Recorder* proclaimed that "there is no longer any possibility of a monopoly of the rubber business as was aimed at by the United States Rubber Company organization."[3] Banigan spent five weeks in the winter of 1896 seeking new customers in Europe, especially Paris and London. Colt was on the Continent at the same time, allegedly searching for capital to purchase the Boston Rubber Shoe Company, fueling the rumors that Colt and Banigan might hook up the two independent operations.[4] Any potential partnership obviously disintegrated during the brutal cross-examination in the salary trial.

"Billious Attack"

When the press originally reported Joseph Banigan's emerging illness, no one knew how sick he really was. The Sisters of Mercy, who administered St. Xavier's Academy in Providence in the heart of the Irish fifth Ward and the site of the Know-Nothing riot in 1855, sent Banigan a spiritual bouquet upon learning of his ailment: They embedded a bronze tablet in the convent's chapel, written in Latin:

> Joseph Banigan, being a distinguished citizen, a most generous patron, and a worthy spouse of Mary, because he has been obliging so many times to his religion and to his country [and] to the fine arts collected, the Chapel of this fair house with the greatest of generosity from his own means they shall build up, they shall adorn.[5]

Banigan, had, after all, donated $25,000 to the convent on Broad Street. After receiving the good wishes of the nuns, he replied in one of only a few extant, personal handwritten letters. He wrote that the sisters' gift brightened one of his rooms and that "the prayers and communions enumerated by the Bouquet for one spiritual benefit are fully appreciated and it is a matter of much gratification to us to be remembered so constantly by those whose prayers carry with them the efficacy of lives lived in the odor of sanctity." Furthermore, he claimed: "I am pleased to say that I feel quite recovered from my illness although I have an occasional reoccurrence of the trouble yet each one is much less painful than the preceding one and I feel encouraged to believe that before long I will be my old self once more."[6] Twenty months later Banigan died.[7]

He underwent a gall bladder operation in New York City in early 1898 by "eminent surgeons," and he returned to Rhode Island to be with his family. A second local medical procedure determined that cancer had permeated his stomach area. Physicians cleared a duct but it was too late. Like the potato blight, the disease spread unchecked. He died on 28 July 1898 at age fifty-nine, his family at his side. He lived longer than most in an era when life expectancy hovered at the fifty-year mark, longevity compromised by childhood diseases. Banigan had escaped the hunger, fevers, and attendant afflictions of the Irish Famine at a young age. He escaped the Irish-American ghetto in Providence, two calamatous accidents, and traveled a scientific and financial trajectory of legendary proportions. He seemed indestructible, but beneath the surface the gastronomical rot took a toll like the spoiled potatoes in Ireland. The ironclad will kept the disease in temporary abeyance, making his death all the more surprising. He looked too vigorous and unprepared to call it a day. The world never stopped for anyone, but Irish Catholics in Rhode Island held their breath momentarily. For good or for bad, the head of the Irish clan had perished. The Church bled. Many of the Irish parishes, beneficiaries of his largesse, openly wept. The rubber king had passed. The uncrowned leader, the point man against innumerable insults heaped upon Irish immigrants, would no longer battle and beat the Yankee adversaries. Irish-Catholics counted another martyr. Joseph Banigan was dead.[8]

Around the time of his illness the union movement almost expired as well. During the interregnum between the Knights of Labor and the rise of the American Federation of Labor, the Catholic Church in Rhode Island and, by implication, other regions in America, supported workers' rights for blue-collar parishioners. Banigan helped his compatriots get a foothold on the ladder to economic success at the rubber company when he hired and trained them during a time of intense discrimination. The Irish boot and shoe makers flourished when given the opportunity, especially in a city like Woonsocket, which contained so many immigrants. The activities of the Knights of Labor in the 1880s, and earlier agitation that embraced the campaign for a ten-hour workday, provided valuable organizational skills to the working class. During the rubber strikes in 1885, local Knights exhibited impressive abilities in public relations, structure, coordination, and a host of other faculties allegedly beyond the scope of mere operatives. Although the union's leadership and segments of the rank and file brought their newfound success to an even higher level, the problems of transforming such groups into vehicles to engage monopolization, overwhelmed them in the short run. The political economy of worker education veered off in another direction for the time being. Locally, the Catholic Church picked up the slack and provided even greater social services beyond the sacristy, as demonstrated in great detail by Professor Evelyn Sterne.[9]

By the time of Banigan's death in 1898, the network of Irish-Catholic organizations that had developed after the Civil War actually proliferated, even as Irish immigration tapered off. Some of the ethnic-oriented organizations came under stricter clerical control while others careened in a more independent direction. Father Michael McCabe, the original guiding light of the faith in the area, ensured before his own death that the St. Charles Christian Doctrine Lyceum still debated controversial subjects in public places. Participants discussed the merits of a city government for Woonsocket, the nature of the Salvation Army, the silver question, and whether America or Ireland produced the greatest orators. The debating group also condemned the chauvinistic American Protective Association, an anti-Irish-Catholic organization that flourished in Rhode Island in the 1890s. Earlier, McCabe displayed some of his own ecumenism by donating to religious and secular causes, in life and death, just as Banigan had.[10]

In 1894 Irish Americans formed a Republican Club in Woonsocket to elect their own kind to the General Assembly, "like the Americans and French-Americans."[11] They may have felt a kinship to the protective policies of the national Republican Party, with so many local industries dependent on foreign raw material, causing the fear of low-wage, job-killing imports. The Irish-dominated Democratic Party seemed mired in marginality. This minor political shift failed to affect the more traditional immigrant activities and culture but provided glimpses of a group about to come of age and assert its political independence, for good or for bad. The fact that Irish Catholics could abandon some customary anchors and embark in a different direction demonstrates a positive, but controversial, move toward acculturation. At the same time the assistant pastor at St. Charles, the Reverend Thomas E. Ryan, formed the Rhode Island Irish Language Revival Society to preserve the vernacular without expectations that the Irish would master it entirely. At a Hibernian night at St. Charles Church in November 1896, all the musical stanzas were in Irish: "The language of the songs in the grand old tongue melted into endearing melody, with a softness and beauty which proved how well this tongue is adapted for bringing out the sentiments of the heart in immortal verse."[12]

The parades on March 17 continued to serve as a barometer for Irish pride during and after Banigan's death: "The day was observed here for many years as a holiday, when it was allowed to pass unobserved even in the city of Providence," and "the green ribbon was everywhere seen."[13] When the holiday fell on a Sunday in 1889, the local newspaper warned that "Descendants of our Puritan fathers need not feel shocked at the announcement that the bells of all the Protestant churches in Woonsocket will be rung on St. Patrick's Day this year."[14] The playful humor reflected, to some degree, the acceptance and assimilation of the Irish in this polyglot city. In 1890, "nearly all the residents" of Millville went to Woonsocket, where most mills closed, to view the procession. Names previously associated with the Knights of Labor often served as marshals for the parades. Joseph McGee, for example, former officer in the Knights, a Woonsocket official, and a Banigan critic, headlined the festivities in 1899.[15]

There seemed to be a greater number of Irish patriotic cultural events in the last years of the Gilded Age as well, like the play *Robert Emmet* presented on St. Patrick's Day in 1888. There were concerts in Woonsocket and Millville, featuring Irish favorites. The Sarsfield Literary Association (which amazingly celebrated the 117th birthday of Robert Emmet in 1895) and the Catholic Ladies Reading Society flourished. A local priest lectured about Celtic literature to the Fortnightly Club. Millville featured the Shamrock Dramatic society. On the Irish nationalist front several groups continued to flourish, including the Irish National League. The Rhode Island branch donated

$7,506.79 (the seventh largest sum in the United States in 1891) and Woonsocket contributed $1,550, one-fifth of the total figure. The League sponsored a Michael Davitt branch in Millville while the *Woonsocket Patriot* described the always militant Clan-na-Gael as advocating "constant drill in the use of firearms." The Ancient Order of Hibernians (AOH) formed a company of Hibernian Rifles, too, and a women's auxiliary.[16] Home Rule advocacy expanded. Patriots and parliamentarians John Stuart, Thomas Gratton, and John Dillon, for example, spoke in Woonsocket, as did many other Irish figures. A local celebrity, James Wilson, an Irish political prisoner who escaped from exile in Australia on the famous ship *Catalpa*, became an international symbol of Fenian resistance. He drew a large crowd at a local rally.[17]

The Father Mathew Temperance Association sponsored meetings and lectures and, at a large celebration in 1899, honored the fiftieth anniversary of the priest's visit to Woonsocket. Banigan probably would have applauded, as he blamed the area's bars for emboldening his workers in 1885. Other good news in Woonsocket at the time included almost 2,000 children attending public school and just over 1,500 enrolling in parochial institutions. The Ancient Order of Hibernians prospered, although relations between the Catholic hierarchy and the group still seemed strained from earlier disagreements, locally and nationally, over such issues as clerical control and independent nationalist activity. Older militant groups like the Land League, now in decline, had set their own agendas.[18]

Novel organizations soon took up the slack. Catholic physicians in Woonsocket started the Crusaders of the Holy Cross at the end of 1890. The city had hosted Banigan, the Knight of St. Gregory, and his employees, organized as the Knights of Labor, while the Catholic Knights of America provided religious, fraternal, and insurance protection. The next fraternal-religious manifestation occurred in 1895 when the Knights of Columbus incorporated and cut across the ethnic divide in Catholic America to include most co-religionists regardless of the country of origin. Thomas H. Murray, the editor of the *Woonsocket Evening Call*, helped form the American-Irish Historical Society in 1897 with a journal dotted with local references. In 1899 the group commemorated the Battle of Rhode Island during the American Revolution and paid tribute to Irish participants, most of whom were Protestant, like General John Sullivan—another reflection of growing maturity. Despite all of this ethnic reaffirmation the Irish still experienced societal problems besides alcoholism, regardless of the efforts by Bishop Harkins and Joseph Banigan to develop social service agencies. Petty crime, poverty-inflicted infectious disease, fighting, and streetwalking continued to afflict the community, although to a lesser degree. Both Yankees and Catholics questioned the role of the latter, from different perspectives, in fighting fellow Catholics in the Spanish American War. Although larger society still doubted Irish-American patriotism and allegiances, St. Charles Church hosted a positive presentation about Americanism to the largest audience in Woonsocket history in the same period.[19]

At times, the Church and individual pastors gingerly followed in Father McCabe's path to foster positive labor-management relations. The Reverend John McCarthy, from the pulpit of the newest Catholic Church in Woonsocket, Sacred Heart (serving the exploding French-Canadian immigrant population), attacked the system of industrial fines whereby textile managers reduced wages for allegedly shoddy workmanship. The priest also pointed out the "condition of servitude" at area factories. Students at Woonsocket High School held a debate that same year as to whether unions were in the best interest of workers, and voted in the affirmative. By the late 1890s the American Federation of Labor recruited skilled workers, including carpenters, masons, plasterers, and others in the building trades, into local union ranks, a far cry from earlier attempts to organize the down-and-out in textile factories around the issue of the ten-hour day. Although the Catholic Church managed to fill the labor void briefly, the radical left also flexed its muscle among disaffected operatives. James P. Reid of Providence, the only socialist ever elected to the Rhode Island legislature, spoke at Monument Square in Woonsocket in 1897. In a strange twist of fate, the rubber workers in Woonsocket turned to the Socialist Labor Party to represent them in 1899 as the Arctic Makers Union. As part of the Socialist Trades and Labor Alliance, the operatives helped plan Woonsocket's first ever Labor Day Parade that year and fielded a socialist ticket in the November elections, garnering about 5 percent of the Woonsocket vote. Money left over from the procession went to form a Central Labor Union in 1900 (still in existence today). In a roundabout fashion, this eventually led to affiliation with the AFL. Banigan's paternalism, the insensitivity of the cartel, and periodic recessions took the progeny of the Knights of Labor from "no conflict with legitimate enterprise—no antagonism to necessary capital" to temporary opposition to the free enterprise system, until the conservative American Federation of Labor eventually gained sway.[20]

Funeral

There were no nostalgic death notices for the old union ways in a society preoccupied with making a dollar faster than ever, regardless of any labor ideology that proclaimed "an injury to one is an injury to all." The obituaries for Joseph Banigan, on the other hand, even in the Yankee press, lauded his remarkable life. On July 30, 1898, two days after his passing, Banigan was laid to rest "with all the beautiful solemnity of the Roman Catholic requiem service."[21] The funeral took place at St. Joseph's Church on the East Side of Providence, the pioneering Irish-Catholic parish of Banigan's youth to which, later in his life, he would challenge the parishioners to match his donations. Now the grizzled veterans of the Rhode Island rubber industry, high and low, kneeled. The clergy also "gathered there in no perfunctory manner, but to pay worthy tributes to one whose unbounded charity had endeared him to them."[22] The rector of Catholic University attended; the nuns of St. Francis of Allegheny, New York, who operated St. Maria's Home paid their respects; and Bishop Harkins genuflected for his financial and spiritual comrade and also delivered the eulogy. Not surprisingly, the prelate lionized Banigan's charitable nature as overshadowing his other accomplishments, "a memorial forever."[23] Four orders of religious sisters attended. Yankee-Republican governor Elisha Dyer, Jr., came. Prominent Irish-Catholic dignitaries served as ushers. Bishop Harkins, in his tribute, made reference to the fact that Banigan's private charities are "known only to Him who knows all things."[24]

Innumerable mourners visited the Banigan mansion off Wayland Square as his body lay in state, including the Little Sisters of the Poor, who managed the Home for the Aged that he financed so long ago, and who would now watch over him from across the railroad tracks in Pawtucket. That charitable endeavor had earned him the knighthood from the Holy See in 1885. Banigan's body traveled to St. Francis Cemetery, where it was interred in the chapel dedicated to his father, the resting place for his extended family. He left $100,000 for its maintenance, although a humanitarian loophole left the interest-swollen endowment open to pillage later. Almost every pallbearer was a veteran colleague from his earliest days in Woonsocket. The clergy included many from his Irish past: Father McCabe's Irish-born brother, a priest in Fall River, Massachusetts; Father Kittredge from St. Augustine's in Millville, who had supported the strikers in 1885; Father Finnegan from Waterbury, Connecticut, Banigan's nephew from Lisirrill; and dozens of others whose connection to the rubber king are lost in a tangle of genealogy and workplaces. Apparently, the only rubber magnate at the service was "Deacon" Converse, the other major holdout with Banigan when the United States Rubber Company formed, although he would soon join the fold after the "rubber king's" departure, as if the death of that independent soul eviscerated the final resister.[25]

Banigan's passing garnered positive obituaries throughout the East, in the Irish-Catholic press, as well as in the country's rubber trade periodicals. The *Providence Journal,* the organ of the state Republican machine and outspoken critic of Irish Catholics, pronounced him one of Rhode Island's richest men, a true accolade from that quarter. Each death notice seemed to mention another financial interest or board of directors associated with the rubber king, some of them previously unknown. Besides the obvious Woonsocket and Providence footwear ventures, he served as president or director of the following companies: American Wringer (Woonsocket), the American Sewed Hand Shoe (Toledo, Ohio), the *Providence Evening Telegram,* Howard Sterling Silversmiths (Providence), Providence Building Company, National Cash Register (Dayton, Ohio), Mosler Safe (Dayton, Ohio), Glenark Knitting (Woonsocket), the Commercial National Bank (Providence), the Industrial Trust Bank (Providence), and Werner Publishing (Utah). He was a major stockholder in the *Atlanta Constitution,* other Utah ventures, the Providence Cable Tramway, and Seamless Rubber (New Haven, Connecticut).[26]

Obituary

Joseph Banigan's biography seems tailor-made for inclusion in the American rags-to-riches pantheon. His Irish-Catholic background, a major barrier to overcome in Yankee Rhode Island at the beginning of his career in the antebellum period, became almost an endearing quality as he grew older, thanks to his ecumenical outlook, indiscriminate charity, and populist character. And with the coming of the "New Immigrants" from eastern and southern Europe in the 1890s, the Irish, a half-century after the Potato Famine exodus, appeared almost Yankee themselves in comparison to the latest wave of peasant newcomers. The Irish brogue, once almost a foreign language to Rhode Islanders who traced their English lineage to colonial times, suddenly seemed like an original American dialect compared to the Latin, Hebrew, Slavic, and Romance languages of the immigrants in the 1890s.

Although Joseph Banigan fervently embraced his Irish-Catholic heritage, his economic success propelled him into the ranks of the local elite. He ingested, almost imperceptibly, the characteristics and lifestyle of those he defeated on the plains of capitalism. In 1877, while still relatively unknown, he rallied the plebian "officers and members of the Irish-American military and civic societies" in Providence for a strawberry festival at Mashpaug Grove on the outskirts of the city. As president of the coalition, he sought to raise funds for the St. Aloysius Orphanage Asylum.[27] A few years later, in 1881, as he achieved prominence, he organized a smaller and different class of his compatriots into the Home Club, a wealthy society for Catholics that would counter the innumerable Protestant establishments that barred ethnic and religious outsiders. Banigan, as the first president, presciently included family members in many of the group's activities. They were not above shucking clams, drinking brandy, and smoking cigars like their Protestant counterparts, but they incisively understood as a minority in Rhode Island the importance of inculcating spouses and children in the trappings of Hibernian cultural success. They purchased a former Providence shore resort and turned it into a worthy competitor to similar Yankee establishments with amenities like bowling alleys, horse riding, and boats. The organization lasted at least until 1895 and may have morphed later into the male Catholic Union established by Bishop Harkins in 1908.[28]

Banigan conscientiously curried the approval of Irish-Catholic toilers and the handful of rising entrepreneurs in his own ethnic community. He hired and trained hundreds of Irish-Catholic workers at the Woonsocket and Millville facilities into skilled male artisans and protected female operatives, thus boosting innumerable working-class Irish immigrants into another economic zone at a time when the slogan "No Irish Need Apply" held sway in the prejudiced industrial corridors of Rhode Island manufacturing. The 1885 strikes, far from being an intemperate orgy of "biting the hand that feeds you," actually demonstrated the tremendous progress that ordinary Irish Americans had made. In that case, Banigan acted more like the English landlord Shirley than his emboldened employees. But though he may have cold-shouldered his striking workers, in combat with his Yankee commercial rivals he demonstrated considerable heat. "It was generally believed that in his business there was additional zest to the competition he encountered because he was an Irish man and a Catholic," wrote the *Providence Visitor*.[29] The rubber king's philanthropic activities did not differ that much from those of enlightened Yankee entrepreneurs, although his ecumenical donations propelled him into the spotlight, if only because Protestants had painted the Irish-Catholic community as narrow and Banigan appeared so large.

As he climbed the ladder of success, Banigan unwittingly joined the Yankee culture. His personal ambition brought him so much success that he seemed like just another native-born stakeholder elbowing for position. While the Knight of St. Gregory oozed Catholic philanthropy from his ever-increasing profit margins, he also summered in Newport, sailed yachts on Narragansett Bay, and built an estate on the East Side of Providence, that rivaled the greatest in the state, supervised by the Gilbane Construction Company. Banigan distributed photographic albums of his Providence home at 510 Angell Street, replete with family oil paintings alongside "a wealth of art treasure from all quarters of the globe."[30] The centerpiece of his collection was Bouguereau's famous creation *Bathing Woman*. The Banigan palace so overwhelmed the affluent neighborhood that the building barely outlasted the owner and was torn down not many years after his death to make way for more affordable but still upscale homes. His earlier, more modest dwelling stood the test of time at 9 Orchard Avenue in Providence. Banigan so completely embraced Yankee ways that local captains of industry elected him president of the almost exclusively Protestant Providence Chamber of Commerce. As far as business was concerned, he was a member of the establishment. The brogue he carried from Ireland, nurtured in the Irish Fifth Ward, probably was undetectable by the end of his life. He conquered his enemies and those of his compatriots, but they had ingested him into their lifestyle. Nonetheless, Ban-

The Banigan Estate on Providence's prestigious East Side about 1890 was comparable to the mansions of Newport. (Author's collection)

igan, a truly Irish representative from the smallest of potato patches, allowed his fellow immigrants to invoke his name and accomplishments against the slings and arrows of a bigot like Wilkinson whose letter to the editor in 1901 challenged the Irish to name any inventor, employer, or scientist from their ranks. They must have shouted "Banigan," rolling the name across their lips.[31]

Banigan's salary trial had mesmerized the rubber industry and, to a lesser degree, other captains of industry. Ironically, the closing arguments were heard in late 1898, after Banigan's passing. Walter Ballou, his right-hand man at Woonsocket, in the U.S. Rubber Company, and finally in the Olneyville venture, had to issue a flyer after his expiration: "The death of our president, Mr. Joseph Banigan, has evidently furnished an opportunity to stock brokers and others manipulating Trust stocks, to again assail this company by circulating false reports to the effect that we have joined or are about to join the Rubber Trust. While we shall greatly miss the wise counsel and advice of our late President, he had placed this company on an independent basis, and there is where we propose to stay."[32] However, in April 1899, the Banigan Rubber Company reorganized. His sons resigned, although the highly experienced Ballou remained. Samuel Colt joined the board preparatory to the takeover by the United States Rubber Company. The news that the rubber king had died eliminated a major rival to the combination and sent the syndicate's stock soaring to a four year high. Some of Banigan's colleagues started a new venture in Woonsocket, the Model Rubber Company. Colt effectively stopped the fragmentation when he orchestrated the purchase of the Banigan Rubber Company and then the Boston Rubber Shoe Company, the last important operation outside monopoly control.[33]

When all was said and done, Joseph Banigan remained a unique and solitary example of unbridled success in Irish-Catholic circles, the largest ethnic group in the state for almost 150 years.[34] By the time of his death, Rhode Island was the first state in the nation with a Catholic majority, even if the Irish, Italians, French-Canadians, Polish, Portuguese, and other Catholic refugees coexisted in an uneasy truce. When that 1967 academic study of the local business elite identified 180 captains of industry, Banigan stood alone among an almost homogenized crowd of Yankee business personalities. The temptation exists to identify Banigan as an example of what the Irish could ac-

United States Rubber Company

Squantum Club
Tuesday, July 1st, 1902

To meet the Superintendents of the Factories of the United States Rubber Company

⸕MENU⸕

Melons

Clam Broth in Cups   Clam Chowder

Broiled Live Lobster

APOLLINARIS
POLAND WATER
MOET & CHANDON
IMPERIAL BRUT

Baked Tautog, Stuffed

Sweet Potatoes   Cucumbers

Frappé Tom and Jerry

Broiled Spring Chicken

CIGARETTES--
NESTOR

Saratoga Potatoes   Green Peas

⸕BAKE⸕

Clams   Clam Fritters   Soft Shell Crabs

Tomatoes Stuffed with Cucumbers

Indian Pudding with Cream

Peaches and Cream   Watermelon

Coffee

CIGARS--SQUANTUM CLUB
LIQUEURS

Three years after Banigan's death, Samuel P. Colt became the president of the United States Rubber Company in 1901. The Superintendents' Club, which Banigan established, met at the exclusive Squantum Club in East Providence in 1902. The waterfront club remains intact. (Author's collection) complish when unshackled from British oppression and imperialism across the Atlantic, as well as what they could achieve through their own talents and ambition when liberated from Yankee domination on this side of the ocean. By the time Banigan died, his countrymen, here and abroad, moved ahead on their own while being pushed from below by global newcomers in the United States. But there would be no other Banigan, although most states seemed to nurture a handful of Irish Americans who ascended the heights of success as he did. Achievement would become a more normal phenomenon accomplished by ordinary Irish Americans no longer driven to prove their mettle at every turn. Banigan had blazed a trail for his fellow Irishmen and Catholics, suffering the pain all pioneers suffer, but accomplishing also a sweet victory in exile from his native land—a victory that carried many others with it.[35]

Kerby Miller, the great chronicler of Irish America, identified openmindedness as one of the modern characteristics possessed by about 20 percent of Irish immigrants when they arrived in the United States between 1800 and 1920. The forward-looking ones, often with some modicum of managerial and commercial experience, came largely from urban areas in Ireland, and exhibited a futuristic outlook free of the hidebound heritage of rural peasants. These newcomers experienced an easier time assimilating into the competitive, fast-paced lifestyle of their adopted homeland. The lion's share of Irish settlers, on the other hand, banded together in urban colonies where the Irish brogue accented the speech even of those born in the United States. These former peasants lacked the burning drive to leave the group and pursue American success. They toiled together almost defensively against a frequently hostile world, in big city settings, often as family units, in cooperative ventures that paralleled the old ways of the Irish countryside. Charles Carroll, a pathbreaking Rhode Island historian, gave his "own kind" a place at the state's head table. He remarked that the patriotic song *The Wearing of the Green* was one of only a handful of that genre that sought a destiny for any group in another nation other than their own, "beyant the say [sea]."[36]

Banigan, of course, by dint of his upbringing, should have belonged squarely in the static category. His life in Lisirrill on the Shirley estate in County Monaghan provided him with a first-class ticket to inertia, according to this justifiably respected interpretation. If rent expense and size of the land meant anything in 1840s Ireland, the Banigans stood last in the townland's socioeconomic order. At the time of the rubber king's death, and occasionally throughout his productive life, commentators would seize on the fact that he was the seventh son of Bernard and Alice Banigan, possessing all the magical power that folklore assigned to that lucky place in the birth order: "to the simple notions of the credulous, a child of superior powers."[37] That analysis points to the difficulty so many Irish experienced in assimilating to life in America; it reminds us of the elderly women who sought out a strand of John Gordon's hanging rope as a talisman. Banigan spent his childhood in an Irish colony, Providence's fifth Ward. When his father, Bernard, died, the patriarch of the family was still listed as a "laborer" in the city directory. Despite her son's wealth, Alice Banigan lived a modest life spent in traditional Irish female activities. Her famous son "was a devout Catholic, possessing the child-like faith which he inherited, above all, from the good old Irish mother whom he all but worshipped."[38] On the surface, Banigan's fate upon his arrival in Rhode Island seemed sealed as tightly as a pair of vulcanized boots.

Seventh son or not, Banigan obviously possessed an intellect capable of breaking out of the fifth Ward fortress that encompassed his adolescent home life and work place. Before he came here, the tiny sliver of land in Ireland did support a "hedge school," an outdoor, clandestine, and primitive attempt to circumvent English restrictions against Irish education. A "class" might literally be held in the camouflaged safety of shrubbery. With only thirty families in Lisirrill, Banigan may have attended some sessions with other children, especially if the operation was seen as a beachhead against English control, although the more enlightened landlord Shirley built a conventional school in that same period. Perhaps the Banigan family's sojourn in Dundee, Scotland—although a failure—exposed him to some other way of life, however briefly. The reckless journey to America itself, while terrifying many sick and forlorn passengers, may have piqued the boy's imagination and fortitude while others huddled in stultifying fear.

Once he arrived in Providence, his job as a child laborer in the neighborhood factory of the New England Screw Company could have teased his fancy rather than dulled his senses. Banigan became such a brilliant businessman that most forgot his mechanical, inventive, and scientific skills honed at a very young age.[39] Perhaps some Yankee supervisor encouraged the young man in the jewelry industry, just as his father-in-law later proctored him in the field of rubber. As a journeyman jeweler Banigan had the skills and wherewithal to travel outside the realm of Providence, away from his own kind. The independent years he spent in Massachusetts with John Haskins, weaned from the nurturing but static confines of the Irish colony, allowed him to tinker and experiment, amassing numerous skills and a fortitude that would come in handy in Woonsocket. After all, he would hold patents and inventions in his early twenties, run a small company, and build a factory that brought him a financial competency seldom enjoyed by Irish-Catholic immigrants. He became a Gilded Age industrial roustabout. Banigan's mechanical skills, combined with his unstinting vision of what could be, probably gained him entry into the era's unofficial scientific fraternity. The ability to get things done and make things happen in the fluid world of experimentation trumped any socioeconomic or ethno-religious disabilities that otherwise might have stained him in Yankee New England. Yet he always kept one foot in the Irish world he could never quite leave behind, a never-ending source of strength to meet the challenges he would encounter in the exclusive lair of Rhode Island Yankees, the world of business. As late as 1893 a piece in the *New York Times* recounted how a reluctant Banigan admitted to building a parish house in Ireland for a priest he befriended. "I'm just running over to the other side of the sea," he told the inquiring reporter, "to attend a house-warming."[40]

Banigan required no special talent to recognize the business dynamics of those scintillating years after the Civil War. Fortunes were made as inventors uncovered the secrets of new products and emerging industries. He also witnessed the loss of wealth by pure scientists who stumbled badly in the world of business. The life of Charles Goodyear, the inventor of vulcanization who was incarcerated for debts, would have been enough of an example. Banigan seemed to have imbibed the requisite ruthlessness to capitalize his inventions without having to experience any such victimization at the hands of preying financiers. If anything, Banigan displayed a merciless and brilliant understanding of legal contracts that continually overwhelmed those who would have taken advantage of him. He gave no quarter in the struggles that took place on such battlefields as municipal tax breaks, footwear competition, or the rubber cartel. In 1885 he lashed out against his own Irish-American employees when they turned on him. Although Banigan narrowly lost that imbroglio, he returned to work with a determination to eliminate the ability of workers ever to hold him over a barrel again. He began the long march to mechanization and the de-skilling of the manual workforce that created his wealth when he founded the automated Marvel Rubber Company. Even without the Knights of Labor strikes, he would have done this eventually. (Ironically, no mention of those walkouts appeared in any death notice or virtually any other subsequent biographical sketch). Banigan obsessively neutralized or consumed every entity along his road to profit that could possibly dip into his potential earnings. He mechanized the work place, purchased suppliers, secured a protected supply of raw rubber, operated an independent distribution network, and hired the most merciless legal practitioners to protect his empire from entrepreneurial pirates. He maintained easy access to necessary capital in Brazil or the United States, and "his name was a familiar one in financial circles in Europe."[41]

While protecting his flank from the robber barons, he developed a global empire through the production of superior goods that engendered positive public relations. He enjoyed further good will by treating his employees respectfully, even after the strikes. Although his celebrated philanthropy probably had many equals among Yankee businessmen, he stood out in the Irish-American community, especially in his home base in Rhode Island, as the champion of his people and their faith. The tandem of Bishop Harkins and Joseph Banigan laboriously enhanced the image of the Catholic Church in the state as an institution of social justice and social services. They would take care of their own. "He was generous to the poor," according to the *Providence Visitor;* "he had been a poor man himself. He willingly shared his wealth with the poor."[42] Banigan's large-scale philanthropy always got advertised whether he liked it or not, although his smaller acts of personal kindness seldom made the front page of any newspaper. In its obituary, the *Woonsocket Reporter,* probably the most knowledgeable source about his career along with the *Woonsocket Patriot,* told how Banigan instructed his private secretary to dispense "a large sum of money to be quietly distributed in such manner that the recipients of his bounty knew not who their benefactor was. This charitable work, which Mr. Banigan directed should be on non-sectarian lines, was particularly effective during the hard times of 1893."[43] He had done the same thing, time and again, during other economic downturns, as in the Panic of 1873. And he paid passages of residents at St. Maria's Home back and forth to Ireland owing to their homesickness or family emergencies.[44]

He liked to talk directly without pretension to his employees, especially when the topic turned to Ireland.[45] "Among his finest personal qualities were his approachableness and his remembrance of friends. He was as unassuming in private as the humblest of his employes [sic], and the men who had worked with him in the shop he never forgot to pleasantly greet or to give sage advice to when that advice was sought," according to the obituary in the *Woonsocket Patriot*.[46] Andrew Donahoe, a rubber worker who died in Woonsocket in 1894, followed Banigan's counsel by amassing $90,000 in company stock, a purchasing policy that Banigan suggested during the strikes of 1885.[47] Another sensitive and incisive obituary from the *Woonsocket Reporter* continued: "Mr. Banigan was a kind, patient friend to young men and he was interested in their progress. He could be keen as the keenest with whom he had business, and he would at the same time lavish the wealth of his experience on some teenager who came to him for advice," further evidence to substantiate his sincere charitable interest in the young.[48] He hired James F. Parker, an apprentice plumber, who did some work at St. Maria's Home in the 1890s. Banigan took a liking to him especially after the teenager refused an offer of full-time work because he said he owed loyalty to the firm that provided his training. Banigan sent word to the contractor that, if he could hire Parker, more subcontracting work would come the proprietor's way. Parker worked for Banigan's family enterprises for decades.[49]

Because of the blistering prejudice against Irish-Catholics that preceded their arrival, the immigrants developed a defensive posture in ethnic colonies, figuratively circling the wagons once they arrived in most parts of the United States. Banigan, to the contrary, made symbolic donations to Protestant, Jewish, and secular causes that broke the mold of perceived Irish intransigence and Catholic narrowness. His tolerance greatly outdistanced his other benefactions. Apparently he partially integrated the construction gangs on his real estate projects in Providence and preached harmony and opportunity for minorities among his staff, high and low, citing the discrimination suffered by the

Irish. He even provided skilled workers and encouragement to the owners of the Outlet Department Store, operated by the Samuel Brothers, two Jewish businessmen who encountered some local prejudice and resistance upon opening their facility in 1894.[50] The *India Rubber World* wrote that "Mr. Banigan was a liberal giver to every religious denomination, deeming those investments as wise as any which paid him no dividends except the blessings of the unfortunate."[51]

Banigan may have lived the life of "lace curtain" ease near the end of his career, but apparently he enjoyed recounting the early days of adversity in private. He liked to shock his listeners by saying, "Nothing that I have touched has ever failed, and my success is as much a surprise to myself as it is to anyone else."[52] The *Providence Visitor,* which Banigan bailed out in 1892 by purchasing its stock, wrote: "It was the talk of those on the outside that the Rubber Trust at the time of its organization would not have shed many tears if it had succeeded in crushing Mr. Banigan."[53] He shared the simple secret of his prosperity in an 1895 address to the Boston Boot and Shoe Club: "A man must give attention to this sort of business, day-in and day-out, seventeen or eighteen hours every day to make a success of it."[54] That formula probably differed little from that of his hardworking Yankee peers, although they had enjoyed a head start. One Rhode Island historian identified the personality traits that Banigan employed in his quest: "Mr. Banigan achieved his phenomenal success solely through his indomitable persistence, business genius, and the constructive imagination which made it possible for him to foresee the place which rubber was destined to take among the industries, and its value as a commodity in everyday life."[55]

Just before his death, the rubber king, propped up in his bed, signed a contract to build a chaplain's home at the Little Sisters of the Poor.[56] A few days later, "he met the end like a Christian. Fortified by the frequent reception of the sacraments, his last days were most edifying and fitly crowned a life which in the midst of the world's temptations to avarice and hardness, and the distractions of worldly care found time and heart to cultivate the virtue of generous almsgiving."[57] Rhode Island historian Thomas Bicknell, author of a mammoth five-volume history of the state in 1920 wrote: "His life was a fine example of what an indomitable will, tireless energy, and inventive genius can achieve in spite of handicaps that dishearten men of less heroic mould."[58]

Postscript

Joseph Banigan left a seven-page will, bequeathing impressive donations to many of the charities he bankrolled during his life. In fact, his philanthropy in death was so weighted in favor of altruism, his grieving wife of twenty-seven years, Maria Conway, contested it. The parties reached an out-of-court settlement that apparently provided her with a greater endowment than initially allocated from the $10,000,000 fortune. Powerful and highly-respected Charles E. Gorman, Banigan's local counsel, and longtime Irish-American activist, represented the estate. Mrs. Banigan received the Angell Street mansion and $15,000 annually before the eventual settlement, but decided to live outside the family homestead. She died prematurely at age fifty-five in 1901 from heart disease. Her funeral paralleled the majesty of her husband's obsequies.[59] She left most of the wealth not already allocated to Banigan's four children from his first marriage to out-of-state brothers, sisters, nieces, and nephews. This dispersal of the fortune outside Rhode Island is probably the reason for the loss of what must have been a mammoth archive of historical material relating to the family and rubber business.[60]

Within a decade after Banigan's death, most of his immediate family followed him to entombment in St. Bernard's Mortuary in Pawtucket's St. Francis Cemetery. No sooner had Maria Banigan passed than William Banigan, Joseph's thirty-six-year-old son died of pneumonia only two weeks after his stepmother. He had attended LaSalle Academy in Providence and Manhattan College before joining his father in the rubber industry at Millville and Woonsocket. He later oversaw the family real estate business. At the time of his death he was a commodore in the elite Rhode Island Yacht Club, having traveled a long way from the family's Irish background. He split his own considerable estate between his wife and daughter.[61] The oldest son, John Banigan, a confidant and associate of his father in the rubber business, died in December 1908 at the age of forty-five.[62] Alice Banigan Sullivan, Banigan's other daughter, and wife of a physician, died in 1909 when she was only forty-three. In 1900 she had commissioned a famous German stained glass company to produce an impressive piece to adorn the altar at St. Joseph's Church in the Fox Point section of Providence in memory of her mother and father, a memorial that sparkles even today. Banigan's oldest sister, Catherine, who had come to Rhode Island during the assisted emigration from County Monaghan, passed in 1906 at the age of seventy-four. His youngest sister, Mary, born in the Irish fifth Ward, lived until 1939, dying at the age of eighty-two. His older brother, Patrick, who married a Protestant, worked occasionally with the rubber master but was not buried in the family chapel. There is no record of the other siblings, who probably died in childbirth in Ireland.[63]

St. Bernard's Mortuary

Banigan had begun construction of his $100,000 mortuary chapel in 1892, although it was not quite finished even when he died. He left another $100,000 in his will for its upkeep. The same local architectural firm that designed St. Maria's Home also did the honors for the intimate chapel Banigan envisioned as a place of spiritual reflection for other visitors to the cemetery. He arranged for a family board of directors, headed by the three oldest members of the immediate family. The founder made a mistake, at least by his standards, in allowing the endowment to be tapped for "family needs" in extreme circumstances. As different generations of the family receded further in time from the founder and his general wealth, the temptation must have been great to employ the last pot of gold for other purposes. Apparently, one director, "Uncle Ned," practically drained the compounded reserve in the late 1920s to start a factory, only to lose it all during the Great Depression.[64]

The Diocese of Providence originally deeded the chapel to Banigan, who deeded it back to the Church before he died. A stipulation in the agreement allowed the diocese to dismantle the structure if it fell into disrepair, despite the initial healthy endowment. As the family walked away from the mortuary years later with no funds left to repair the fraying structure, Bishop Russell J. McVinney tore down the original sepulcher and replaced it in the early 1970s with a bland, almost hideous holding tomb. McVinney's own resting place at *Regina Cleri* Cemetery at Our Lady of Providence Seminary in Warwick was relocated to St. Francis in 1986. His family is buried on the grounds there which were originally known as the Bishop's Cemetery. Whatever the circumstances, the Catholic Church failed to respect the memory of

A postcard showing Banigan's family mortuary at St. Francis Cemetery in Pawtucket built at a cost of $100,000 in 1898. When a family member was able legally to withdraw maintenance funds, the chapel fell into disrepair, and the Catholic Church razed the structure. (Postcard, author's collection)

Bishop Russell J. McVinney replaced it with a virtual concrete bunker in the 1970s although the family mortuary exists beneath it. (Photo of the replaced mortuary by Phil Banigan)

one of its greatest benefactors by not rehabilitating the mortuary, regardless of the circumstances.[65]

Full Circle

While Joseph Banigan's casket lay in state at his home before the funeral, a mysterious, unknown young man stood almost at attention for hours gazing at the "rubber king." Few visitors recognized the out-of-state college student. Frank Byrne, his father, had allegedly masterminded the assassination in Dublin in May 1882 of the English Lord Frederick Cavendish, the Chief Secretary for Ireland, by a select group of Fenians called the Invincibles. Although several conspirators hanged or went to prison, Byrne escaped to France (where he served meritoriously in the Franco-Prussian War of 1870–1871) and avoided extradition before leaving for New York. He had been the chief secretary to Charles Stewart Parnell. Around 1890, Byrne moved to Providence and worked at the Hanley Brewery with several other Irishmen wanted by British authorities. When the revolutionary died in 1894, local newspapers took a great interest in whether Byrne had actually been the chief conspirator. Byrne was buried in St. Mary's Cemetery in Pawtucket, ironically on the land provided by the once liberal Wilkinson family. John Gordon, the last person executed by the state of Rhode Island in 1845, probably was interned there too. In 1899, 1,000 Irish Americans gathered at the graveyard to dedicate a monument to Byrne.[66]

While on his death bed in 1894, Byrne asked Capt. Joseph Mullen, a State House political operative, to arrange for the care and education of his son. Mullen convinced his close friend Joseph Banigan to assist, "with the result that a good home and exceptional educational advantages had been given him through the generosity of the large-hearted man who knew no creed or color when dispensing his charities."[67] The student had never met his benefactor in life. Now he paid his respects in death. Banigan, who would never have countenanced the elder Byrne's violent activities, still picked up the tab for the son of the Irish insurrectionist. The long journey from Lisirrill had come full circle. Joseph Banigan paid his dues and those of many other Irish immigrants. He helped his countrymen of all stripes come of age in the "land beyant the say."[68]

NOTES

INTRODUCTION: SPOUTING (pages 1–10)

1. *Providence Journal,* 3 March 1901, (hereafter *PJ*).
2. Patrick T. Conley and Matthew J. Smith, *Catholicism in Rhode Island: The Formative Era* (Providence: Diocese of Providence, 1976), 25; Robert W. Hayman, *Catholicism in Rhode Island and the Diocese of Providence, 1780–1886,* 2 vols. (Providence: Diocese of Providence, 1982), 2:21–22; Gary Kulik, "Pawtucket Village and the Strike of 1824: The Origins of Class Conflict in Rhode Island," *Radical History Review* 17 (spring, 1978), 5–37.
3. Patrick T. Conley, *The Irish in Rhode Island: A Historical Appreciation* (Providence: Rhode Island Heritage Commission and Rhode Island Publications Society, 1986).
4. Conley, *Irish in Rhode Island,* 18–19; Mary Josephine Bannon, ed., *Autobiographic Memoirs of Patrick J. McCarthy* (Providence: Providence Visitor Press, 1927); Robert A. Wheeler, "Fifth Ward Irish: Immigrant Mobility in Providence, 1850–1870," *Rhode Island History* 32 (1973), 53–61.
5. The tercentenary of Benjamin Franklin (1706–2006) occasioned even more biographies in an already choked bibliography. In 2004, Yale University published the thirty-seventh volume of Franklin's papers, an 882-page tome edited by Ellen R. Cohn that covered his life for only five months, from 16 March through 15 August 1782. In 2005, after a challenging thirty-year publishing period under several editors, the Rhode Island Historical Society and the University of North Carolina Press completed an amazing thirteen-volume series, totaling more than 8,000 pages of writings and correspondence to and from Rhode Island Revolutionary War General Nathanael Greene. This treasure trove is all the more remarkable inasmuch as Greene died at the early age of forty. Any scholar writing about this officer could stack the volumes next to a writing table and not have to travel much farther to compose a rich biography. For the other side of the academic divide where primary sources are scarce, see the masterful history of a free black man based on limited and confusing direct primary material, Nick Salvatore, *We All Got History: The Memory Books of Amos Webber* (New York: Random House, 1996); or Laurel Thatcher Ulrich, *A Midwife's Tale: The Life of Martha Ballard, Based on Her Diary, 1785–1812* (New York: Vintage Books, 1990), which transforms a minimalist diary into a compelling story.
6. The quotations appeared in his first opinion piece, compiled in a pamphlet, *Letters on Irish Emigration* (Boston: Phillips, Sampson and Co., 1852), 6.
7. William G. McLoughlin, *Rhode Island: A Bicentennial History* (New York: Norton, 1978); Scott Molloy, Carl Gersuny, and Robert Macieski, *Peacably if we*

can. *Forcibly if we must! Writings By and About Seth Luther* (Kingston: Rhode Island Labor History Society, 1998).

8. Patrick T. Conley, *Democracy in Decline: Rhode Island's Constitutional Development, 1776–1841* (Providence: Rhode Island Historical Society, 1977); George M. Dennison, *The Dorr War: Republicanism on Trial, 1831–1861* (Lexington: University Press of Kentucky, 1976); Marvin E. Gettleman, *The Dorr Rebellion: A Study in American Radicalism, 1833–1849* (New York: Random House, 1973); Conley, *Irish in Rhode Island.*

9. Ian S. Haberman, "The Rhode Island Business Elite, 1895–1905: A Collective Portrait," *Rhode Island History* 26 (1967), 38; Scott Molloy, "The Irish in Rhode Island: A long struggle to enter the mainstream," *PJ*, 17 March 1997.

10. Cited in a letter to the editor by Robert Eastman, *PJ*, 26 March 1899; another letter supported the Irish, *PJ*, 28 July 1901.

11. *Woonsocket Patriot*, 8 May 1868, (hereafter *WP*); James W. Smythe, *History of the Catholic Church in Woonsocket and Vicinity, from the Celebration of the First Mass in 1828 to the Present Time, With a Condensed Account of the Early History of the Church in the United States*, (Woonsocket: Charles E. Cook, Printer, 1903), 84–89; Hayman, *Catholicism in Rhode Island*, 1:102–19.

12. For an example, see Glen Porter, *The Rise of Big Business, 1860–1920* (Wheeling, Ill.: Harlan Davidson, 2006, 3rd ed.).

13. Timothy J. Meagher, *Inventing Irish America: Generation, Class, and Ethnic Identity in a New England City, 1880–1920* (Notre Dame: University of Notre Dame Press, 2001), 5, 10. The entire introduction is an excellent historiographical essay on immigration in general and Irish America in particular; J. Matthew Gallman, *Receiving Erin's Children: Philadelphia, Liverpool, and the Irish Famine Migration, 1845–1855* (Chapel Hill: University of North Carolina Press, 2000), 5.

CHAPTER 1: THE IRISH BACKGROUND (pages 11–28)

1. This chapter presents a brief overview of Irish history during the childhood of Joseph Banigan. The number of books and articles about Irish history, the Famine, and the subsequent emigration, especially to the United States, is prodigious and overwhelming. I have tried to offer a snapshot of that world, particularly in County Monaghan, displaying some broad contours of how the family might have experienced those events, although there are only a couple of actual primary sources relating to the Banigans during those years. One researcher told me I was fortunate to find anything at all about such marginal lives in that era. Except for placing the Banigans in the firmament of the times, this chapter makes no pretense at original research.

2. Kerby Miller, *Emigrants and Exiles: Ireland and the Irish Exodus to North America* (New York: Oxford University Press, 1985), introduction. Miller's monumental book is the starting place for understanding Irish America and includes a voluminous transatlantic bibliography of manuscript resources that reinforce his thesis of Irish expulsion.

3. Banigan's experience in these areas will be explored in subsequent chapters.

4. Karl S. Bottigheimer, *Ireland and the Irish: A Short History* (New York: Columbia University Press, 1982), 142. Miller (*Emigrants and Exiles*, 42) estimated that just 10,000 Protestant families owned Ireland. See Miller's categories for the size of farms, 48–54.

5. Thomas E. Hachey, Joseph M. Hernon Jr., and Lawrence J. McCaffrey, *The Irish Experience: A Concise History* (New York: M. E. Sharpe, 1960), 53–55; Miller, *Emigrants and Exiles*, 21–22.

6. Peadar Livingstone, *The Monaghan Story: A Documented History of the County Monaghan from the Earliest Times to 1976* (Enniskillen: Clogher Historical Society, 1980), 147. The Clogher Historical Society, publisher of this exhaustive tome, is a highly professional organization that manages to balance popular history with academic investigation, especially in its impressive annual journal.

7. Bottigheimer, *Ireland and the Irish*, ch. 5. For an incisive critique of the Penal Laws, see the older but still valuable Seumas MacManus, *The Story of the Irish Race: A Popular History of Ireland* (New York: Devin-Adair Company, [1921]), ch. 53, "The Later Penal Laws." Kerby Miller attributed some Irish fatalism to the acts, in *Emigrants and Exiles*, 21–22.

8. Cecil Woodham-Smith, *The Great Hunger: Ireland, 1845–1849* (New York: Signet Books, 1962), 21–22.

9. Livingstone, *The Monaghan Story*, 189–90. For the Rhode Island story, see Miller, *Emigrants and Exiles*, 46; Hachey et. al, *The Irish Experience*, 53–55, 81–86; Mary E. Daly, *The Famine in Ireland* (Dublin: Dublin Historical Association, 1986), 47.

10. The quotation is from the title of the twenty-first installment of a series of articles about the Famine written in 2004 for the *Northern Standard* newspaper in County Monaghan. Shirley uttered the statement at a meeting of the board of guardians of a workhouse in Carrickmacross in November 1849.

11. Hachey et al., *The Irish Experience*, 92–93; the concise history is too concise in this instance, dedicating a mere two pages to what is arguably a momentous event in Irish history; Livingstone, *The Monaghan Story*, ch. 3; Christine Kinealy, *A Death-Dealing Famine: The Great Hunger in Ireland* (Chicago: Pluto Press, 1997), ch. 2. Interestingly, a popular and patriotic children's history of Ireland by MP A. M. Sullivan, originally published in 1867, was reprinted in Providence in 1885: *The Story of Ireland; A Narrative of Irish History, From the Earliest Ages to the Insurrection of 1867, Written for the Youth of Ireland* (Providence: Henry McElroy, 1885). Its few pages on the Famine, 554–59, are truculent. See also James S. Donnelly Jr., *The Great Irish Potato Famine* (Gloucestershire: Sutton Publishing, 2001).

12. Woodham-Smith, *The Great Hunger*, ch. 5; Daly, *Famine in Ireland*, 53; Livingstone, *The Monaghan Story*, 216; Brian MacDonald, "A Time of Desolation:

Clones Poor Law Union, 1845–50," *Clogher Record* (Journal of the Clogher Historical Society) 17:1 (2000), 15. The *Clogher Record* comprised two large volumes for 2000 and 2001, packed with primary source material for the area around County Monaghan during the Famine, an absolute treasure trove, especially for the foreign researcher.

13. "Evictions and disease add to the plight of the poor," *Northern Standard*, part 11, 6 May 2004.

14. Livingstone, *The Monaghan Story*, 217–21; Daly, *Famine in Ireland*, 100–4; Donald M. MacRaild, ed., *The Great Famine and Beyond: Irish Migrants in Britain in the Nineteenth and Twentieth Centuries* (Dublin: Irish Academic Press, 2000), introduction, 2; "Evictions and disease . . ." part 11. In 1841 the counties that make up the Republic of Ireland today had a population of 6.5 million. That number dwindled to a low point of 2.8 million inhabitants by 1961, and by 2005 the numbers in the 26 counties had climbed to just over 4 million. Robert P. Connolly, "The New Face of Ireland: More Hellos than Goodbys," *Boston Irish Reporter*, Nov. 2005.

15. John Mitchel, *The Last Conquest* (New York: n.p., 1860), cited in Miller, *Emigrants and Exiles*, 306.

16. Kinealy, *Death-Dealing Famine*, ch. 1; Daly, *Famine in Ireland*, 63–66; Miller, *Emigrants and Exiles*, 305–306. Researchers at the University of Wisconsin–Madison recently discovered a gene from a wild potato that, when spliced into the code of commercial potatoes, creates a resistance to the blight, *PJ*, 15 July 2003.

17. Livingstone, *The Monaghan Story*, 95, 114, 134; Miller, *Emigrants and Exiles*, ch. 2.

18. Livingstone, *The Monaghan Story*, 138–42, 177–82; Patrick Duffy, "The Famine in County Monaghan," in Christine Kinealy and Trevor Parkhill, eds., *The Famine in Ulster: The Regional Impact* (Belfast: Ulster Historical Foundation, 1997), 171; Kinealy and Parkhill, introduction, 3; Miller, *Emigrants and Exiles*, 39, For the violence in that era, see Miller, 60–66; Edward T. McCarron, "Altered States: Tyrone Migration to Providence, Rhode Island During the Nineteenth Century," *Clogher Record* 16:1 (1997), 148–49.

19. Livingstone, *The Monaghan Story*, 184–210.

20. The Public Records Office of Northern Ireland (PRONI), located in Belfast, holds the massive and detailed records of the Shirley Estate. There is a finding guide with a good but unpaginated introduction. PRONI: D3531/R/6/11, p. 108; D3531/R/7/2; D3531/S/58; D3531/S/56/100. In most of the biographical sketches of Banigan in American sources, his birthplace is listed and spelled as "Carrickadoncey" in County Monaghan. No such place exists, but the neighboring townland to Lisirrill is Carickadooey, where several Banigans did reside. In fact, the maiden name of Banigan's mother, Alice, was also Banigan and perhaps she hailed from there; but the name became bastardized, as so often happened to immigrants. Even the name Lisirrill differs in spelling in the ordinance survey, rent roll, and other sources; I

have used the spelling by Patrick J. Duffy, *Landscapes of South Ulster: A Parish Atlas of the Diocese of Clogher* (Antrim: Institute of Irish Studies of the Queens University of Belfast in association with the Clogher Historical Society, 1993).

21. Duffy, "Famine," 171; Livingstone, *The Monaghan Story*, 211.

22. Duffy, "Famine," 171.

23. Duffy, "Famine," 170–77; Kinealy, *Death-Dealing Famine*, ch. 6; Daly, *Famine in Ireland*, 109; Patrick Duffy, "Management Problems on A Large Estate in Mid-Nineteenth Century Ireland: William Steuart Trench's Report on the Shirley Estate in 1843," *Clogher Record* 16:1 (1997), 101; There is a discrepancy in the measurements of the estate depending on the determination of Irish or English acres, the latter being almost double in size, Duffy, "Remarks on Viewing the Estate of John Shirley Esquire Situated at Carrickmacross, In the Barony of Farney in the County of Monaghan, Ireland," *Clogher Record* 12:3 (1987), 300.

24. Livingstone, *The Monaghan Story*, 199; Ruth Ann Harris, "Negotiating: Irish Women, the Landlord, and Emigration from the Shirley Estate, Carrickmacross, Co. Monaghan," unpub. manuscript, Boston College, 1, 4, 9–10; Duffy, "Famine," 181–82. Paddy Duffy, "Colonial Spaces and Sites of Resistance: Landed Estates in 19th Century Ireland," in Lindsay J. Proudfoot and Michael M. Roche, *(Dis)Placing Empire: Renegotiating British Colonial Geographies* (London: Ashgate, 2005), 33.

25. William Steuart Trench, *Realities of Irish Life* (1868) (London: Macgibbon and Kee, 1966), 15–16.

26. Trench, *Realities*, 21; Livingstone, *The Monaghan Story*, 213; Patrick Duffy, "Assisted Emigration," *Clogher Record* 14 (1992), 11; Duffy, "Management Problems," 109. In the Rhode Island History courses I teach, I require students to write a short genealogical paper. In one recent account a student recalled how his Irish-Catholic ancestors tended a farm in Connecticut in the early twentieth century. Several family members built subsequent homes on the periphery of the property, Brandon Pearce, "A Rhode Island Swamp Yankee," undergraduate unpub. paper, University of Rhode Island, 3.

27. Trench, *Realities*, 25; Livingstone, *The Monaghan Story*, 214; Duffy, "Assisted Emigration," 11.

28. Trench, *Realities*, 29.

29. Trench, *Realities*, 28–29; "Thousands protest at high rents on Shirley Estate," *Northern Standard*, part 2, 19 Feb. 2004; L. O. Mearain, "Estate Agents in Farney: Trench and Mitchell," *Clogher Record* 10:3 (1981), 407. (The article includes an eyewitness report by a local priest.) "Finding Guide, Shirley Estate," n.p., PRONI.

30. Evelyn Philip Shirley, Esquire, *The History of the County of Monaghan* (London: Pickering and Co., 1879), preface, n.p.

31. Trench, *Realities*, 30–38; Livingstone, *The Monaghan Story*, 127, 196, 202–203, 207, 214; Duffy, "Assisted Emigration," 13.

32. See chapter 2 herein.

33. Patrick Kavanagh, introduction, *Realities of Irish Life,* 11; Patrick J. Duffy, "'Disencumbering Our Crowded Places': Theory and Practice of Estate Emigration Schemes in Mid-Nineteenth Century Ireland," in Patrick J. Duffy, ed., *To and From Ireland: Planned Migration Schemes, c. 1600–2000,* 88.

34. Trench, *Realities,* 39–43; Duffy, "Assisted Emigration," 14; Duffy, "Famine," 177, 183, *Clogher Record,* 17:1 (2000), 13; Duffy, "Management Problems," 101–22.

35. Angelicque Day and Patrick McWilliams, *Ordinance Survey Memoirs of Ireland: Counties of South Ulster, 1834–8* (Belfast: Institute of Irish Studies, 1998), 141.

36. Duffy, "Famine," 184.

37. *Clogher Record* 18:1 (2000), 87; Daly, *Famine in Ireland,* 67–69; Kerby Miller, "Revenge for Skibbereen," in Arthur Gribben, ed. *The Great Famine and the Irish Diaspora in America* (Amherst: University of Massachusetts Press, 1999), 181–82, 186; Kinealy and Parkhill, eds., *Famine in Ulster,* introduction, 1–2; Livingstone, *The Monoghan Story,* 217.

38. Trench's report is in PRONI, D.3531/S/55, cited in Duffy, "Famine" 181. Kinealy and Parkhill, introduction, 1–2; Duffy, "Famine," 169–71, 178–79, 181. For specific examples of the land question as it related to women, see the Harris article, "Negotiating." Day and McWilliams, *Ordinance Survey,* 140; Duffy, "Assisted Emigration," 12; Livingstone, *The Monaghan Story,* 222; *Clogher Record* 17:1 (2000), 1.

39. Trench, *Realities,* 47.

40. *Clogher Record* 17:1 (2000), 35, 335.

41. *Northern Star,* 1 Aug. 1846, cited in *Clogher Record* 17:1 (2000), 33.

42. *Clogher Record* 17:1 (2000), 52, 188, 335; *Clogher Record* 17:2 (2001), 479; Daly, *Famine in Ireland,* 44–47, 70–71, 94; Duffy, "Assisted Emigration," 15; Livingstone, *The Monaghan Story,* 286; Harris, 23.

43. *Clogher Record* 17:1 (2000), 331–32, 479.

44. *Clogher Record* 17:2 (2001), 448.

45. Trench, *Realities,* 51.

46. Duffy, "Famine," 185, 194; *Clogher Record* 17:1 (2000), 83–85; Kinealy, *Death-Dealing Famine,* chs. 4–5; Daly, *Famine in Ireland,* 92; Livingstone, *The Monaghan Story,* 211.

47. Cited in Kinealy, *Death-Dealing Famine,* 148.

48. Duffy, "Assisted Emigration," 11–3, 16; Duffy, "Famine," 193; Daly, *Famine in Ireland,* 108, 120; Harris, "Negotiating," 3, 7–8; Kinealey, *Death-Dealing Famine,* ch. 6; Day and McWilliams, *Ordinance Survey,* 138; Livingstone, *The Monaghan Story,* 222; PRONI, Shirley Estate, D3531/R/6/13, D3531/P/ Box 1, Emigration Papers. Gerard Moran, "'Shovelling out the Poor': Assisted Emigration from Ireland from the Great Famine to the Fall of Parnell," in Duffy, ed., *To and From Ireland.* Moran estimated that the number of emigrants receiving help was higher than the usually accepted figure of 5 percent of passengers, 139. Duffy, *Landscapes,* 17. Duffy, "Colonial Spaces," 33.

49. *Clogher Record* 17:1 (2000), 102.

50. Duffy, "Assisted Emigration," 24, 52; Duffy, "Famine," 188; *Clogher Record* 17:1 (2000), introduction, 60, 102; Daly, *Famine in Ireland*, 36, 107–108; PRONI, Shirley Estate, D3531/P/Box 1, Emigration Papers; Banigan obituary, *New York Times*, 29 July 1898; for the experience in Liverpool for putative emigrants, J. Matthew Gallman, *Receiving Erin's Children: Philadelphia, Liverpool, and the Irish Famine Migration, 1845–1855* (Chapel Hill, University of North Carolina Press, 2000).

51. Daly, *Famine in Ireland*, 108; Terry Coleman, *Passage to America: A History of Emigrants from Great Britain and Ireland to America in the Mid-Nineteenth Century* (London: Penguin Books, 1974); Jim Rees, *Surplus People: The Fitzwilliam Clearances 1847–1856* (Cork: The Collins Press, 2000), ch. 6, for a detailed look at the passage to North America. James J. Mangan, ed., *Robert Whyte's 1847 Famine Ship Diary* (Dublin: Mercier Press, 1994), written by a Protestant.

52. Daly, *Famine in Ireland*, 117.

53. Kinealy, *Death-Dealing Famine*, epilogue, "The Famine Killed Everything"; Christine Kinealy, "The Great Irish Famine—A Dangerous Memory?" in Gribben, ed., *The Great Famine*, 239–53; A controversial and provocative examination of race, ethnicity, and class is Noel Ignatiev, *How the Irish Became White* (New York: Routledge, 1995); Catherine M. Eagan, "'White,' If 'Not Quite'": Irish Whiteness in the Nineteenth-Century Irish-American Novel," in Kevin Kenny, *New Directions in Irish-American History* (Madison: University of Wisconsin Press, 2003).

CHAPTER 2: RHODE ISLAND'S YANKEE ASCENDANCY (pages 29–58)

1. Willaim G. McLoughlin, *Rhode Island: A Bicentennial History* (New York: Norton, 1978), ch 1; Sydney V. James, *Colonial Rhode Island: A History* (New York: Charles Scribner's Sons, 1975), chs. 1–2; Sydney V. James, *The Colonial Metamorphoses in Rhode Island: A Study of Institutions in Change* (Hanover: University Press of New England, 2000), chs. 1–2; Bruce C. Daniels, *Dissent and Conformity on Narragansett Bay: The Colonial Rhode Island Town* (Middletown: Wesleyan University Press, 1983), chs. 1–3.

2. McLoughlin, *Rhode Island*, chs. 1–2; Timothy L. Hall, *Separating Church and State: Roger Williams and Religious Liberty* (Urbana: University of Chicago Press, 1998), chs. 1–3.

3. McLoughlin, *Rhode Island*, 35–37; Barry D. Mowell, "Massachusetts," in Burt Feintuck and David H. Watters, eds., *The Encyclopedia of New England* (New Haven: Yale University Press, 2005), 694–95; Kerby Miller, *Emigrants and Exiles: Ireland and the Irish Exodus to North America* (New York: Oxford University Press, 1985), ch. 1. In 1719 an unratified law appeared in the colony's laws that denied toleration to Catholics; it was eliminated after the Revolution and probably had little impact, as Catholic immigration was still in the future, McLoughlin, *Rhode Island*, 74–75.

4. Conley, *Democracy in Decline: Rhode Island's Constitutional Development, 1776–1841* (Providence: Rhode Island Historical Society, 1977), chs. 1–2; McLoughlin, *Rhode Island,* 38–39; James, *Colonial Rhode Island,* 67–72; Joyce Botelho, *Right and Might: The Dorr Rebellion and the Struggle for Equal Rights,* 4 pamphlets (Providence: Rhode Island Historical Society, 1992), 1:13–14. These excellent booklets accompanied an exhibition at the Rhode Island Historical Society on the 150th anniversary of the Dorr War in 1992 and included important graphics and illustrations from the period. McLoughlin, *Rhode Island,* 39.

5. McLoughlin, *Rhode Island,* ch. 3; Florence Parker Simister, *The Fire's Center: Rhode Island in the Revolutionary Era, 1763–1790* (Providence: Rhode Island Bicentennial Foundation, 1979); Patrick T. Conley, "The Battle of Rhode Island, 29 August 1778: A Victory for the Patriots," *Rhode Island History* 62:3 (fall, 2004), 51–65.

6. Barbara M. Tucker, *Samuel Slater and the Origins of the American Textile Industry, 1790–1860* (Ithaca: Cornell University Press, 1984); Gary Kulik, "Pawtucket Village and the Strike of 1824: The Origins of Class Conflict in Rhode Island," *Radical History Review* 17 (1978), 5–37. Peter J. Coleman, *The Transformation of Rhode Island, 1790–1860* (Providence: Brown University Press, 1969), ch. 3.

7. Kulik, "Pawtucket Village."

8. Conley, *Democracy in Decline,* chs. 6–8; Botelho, *Right and Might,* 1:14–23. See the bristling attack on Irish-Catholic, working-class enfranchisement by state representative Benjamin Hazard, *Report on the Committee on the Subject of an Extension of Suffrage* (Providence, 1829).

9. Patrick T. Conley and Matthew J. Smith, *Catholicism in Rhode Island: The Formative Era* (Providence: Diocese of Providence, 1976), 39.

10. Charles Carroll, *Rhode Island: Three Centuries of a Democracy,* 4 vols. (New York: Lewis Historical Publishing, 1932), 2:1149; Mary Nelson Tanner, "The Middle Years of the Anthony-Brayton Alliance; or Politics in the Post Office, 1874–1880," *Rhode Island History* 22 (1963), 65–76; Robert Dunne, *Antebellum Irish Immigration and Emerging Ideologies of "America": A Protestant Backlash* (New York: The Edward Mellen Press, 2002).

11. Robert W. Hayman, *Catholicism in Rhode Island and the Diocese of Providence, 1780–1886,* 2 vols. (Providence: Diocese of Providence, 1982), 1:12–18; Miller, *Emigrants and Exiles,* ch. 6. Most Irish immigrants before 1800 were Protestants from Ulster, Maurice J. Bric, "Patterns of Irish Immigration to America, 1783–1800," in Kevin Kenny, ed., *New Directions in Irish-American History* (Madison: University of Wisconsin Press, 2003).

12. The incomplete letters and papers are in a slender folder of the law firm of Richard Ward Greene at the Rhode Island Historical Society Library. Scott Molloy, "Brave Pre-Famine Irish Immigrants," *Providence Journal* (hereafter *PJ*), 17 March 2004.

13. Molloy, ibid. Local Irish immigrants found a more dependable group of

ticket agents by the 1840s, Edward T. McCarron, "Altered States: Tyrone Migration to Providence, Rhode Island during the Nineteenth Century," *Clogher Record* 16:1 (1997), 150; Joseph W. Sullivan, "Reconstructing the Olney's Lane Riot: Another Look at Race and Class in Jacksonian Rhode Island," *Rhose Island History* 65 (summer 2007), 50–51.

14. Patrick T. Conley, *The Irish in Rhode Island: A Historical Appreciation* (Providence: Rhode Island Heritage Commission and Rhode Island Publications Society, 1986), 7–8; Hayman, *Catholicism in Rhode Island*, 1:12–18; Evelyn Savidge Sterne, *Ballots and Bibles: Ethnic Politics and the Catholic Church in Providence* (Ithaca: Cornell University Press, 2004), 36–43; Gail Fowler Mohanty, "'All Other Inventions Were Thrown into the Shade': The Power Loom in Rhode Island," in Douglas M. Reynolds and Marjory Myers, eds., *Working in the Blackstone Valley: Exploring the Heritage of Industrialization* (Woonsocket: Rhode Island Labor History Society, 1991), 69–88; *Republican Herald*, 20 March 1839; Molloy, "Brave Pre-Famine Immigrants." Robert Hayman unselfishly shared the first two manuscript chapters of a planned history: "The Anchor and Shamrock: The Irish in Rhode Island," which sketched pre-Famine immigrants in the state and includes an important collection of federal era newspaper clippings about the subject. Cited as Hayman, MS; ch. 1, 27–28.

15. *Providence Patriot*, 1 Sept. 1824; *Rhode Island American and Providence Gazette*, 31 Jan. 1826, 19 April, 16, 19 Sept. 1828; *Newport Mercury*, 16 Sept. 1826, 3 March, 14 July, 6 Oct. 1827, 4 July 1828, 29 Aug. 1829, 31 Aug. 1833, 22 March 1834, 10 Sept., 12 Nov. 24 Dec. 1836, 28 Oct., 9 Dec. 1837, 5 July 1839; *Columbian Phenix*, 3 April 1827, 22 March 1828; *PJ*, 26 Dec. 1836, 14 Aug. 1840; *Pawtucket Chronicle and Register*, 19 May, 9 June 1837. Some of these citations are included in Hayman, *Catholicism in Rhode Island*, ch. 1, "The Beginnings of Nineteenth Century Irish Immigration." The quotation in the last line is from John Brown Francis to Elisha Potter, 23 July 1850, *Francis Papers*, Series 1, Box 2, Folder 2, RI Hist. Soc. Lib. The gravestone inscriptions are reproduced in "Tombstones," in the guide to the play *A State of Hope*, n.a., n.d., n.p.; McCarron, "Altered States," 159.

16. *PJ*, 16 Nov. 1839.

17. *PJ*, 16 Nov., 21 Dec. 1839. Reports of the trial appeared in the last week of November 1839.

18. James W. Smyth, *History of the Catholic Church in Woonsocket and Vicinity from the Celebration of the First Mass in 1828 to the Present Time* (Woonsocket: Charles E. Cook, Printer, 1903), ch. 9. Hayman, *Catholicism in Rhode Island*, 1:24, 28. Irish canal workers had a reputation of being belligerent, but a smaller, mixed workforce and competent management mitigated any local upheavals. Peter Way, *Common Laborers: Workers, and the Digging of North American Canals, 1780–1860* (Baltimore, Johns Hopkins University Press, 1993), 63, 93–99; Matthew E. Mason, "'The Hands Here Are Disposed to Be Turbulent': Unrest Among the Irish Trackmen of the Baltimore and

Ohio Railroad, 1829–1851," *Labor History*, 39:3 (1998). *Woonsocket Patriot* (hereafter *WP*), 4 July 1878.

19. *WP*, 9 Oct. 1874, 4 July 1878, 17 Feb. 1893, 17 Nov. 1899. Interestingly, and offering a preview of Woonsocket's ethnic and religious diversity, a Protestant minister introduced Fr. Theobald Mathew at the meeting. Hayman, MS, ch. 2, 28–29. An excellent account of the priest's career is John F. Quinn, *Father Mathew's Crusade: Temperance in Nineteenth-Century Ireland and Irish America* (Boston: University of Massachusetts Press, 2002); Patrick Reddy, Michael's son, was born in Woonsocket in 1839 and attended public school. In 1860 he left for California to dabble in gold mining, eventually becoming a lawyer in San Francisco who championed the causes of "the working classes." Elected to the California senate, he often took high-profile legal cases like the successful defense of militant union miners in Coeur d'Alene, Idaho, in 1892. Reddy's triumphant visit to his hometown in 1893 triggered memories of his beloved father in the local press and the arduous Irish road to success. *WP.*, 17 Feb. 1893; *PJ*, 12 Oct. 2003.

20. *Providence Sunday Tribune*, 5 March 1911; Hayman, *Catholicism in Rhode Island*, 1:37. See also Hayman, MS, ch. 1, pp. 13, 25, for earlier mention of several other bleachers coming to Rhode Island from County Monaghan.

21. One newspaper provided a view of a utopian existence, *Providence Sunday Tribune*, 5 March 1911, while another included instances of prejudice, *Providence Sunday Journal*, 14 Sept. 1919. The sense of exile and loss reflected Kerby Miller's observations about the Famine refugees, Miller, *Emigrants and Exiles*, introduction. Hayman, MS, ch. 2, p. 8.

22. Carroll, *Rhode Island*, 2:1150; one immigrant from Ballybay who arrived in 1852 progressed to the position of overseer at the Bartlett Mill in Woonsocket, *WP*, 22 April 1881.

23. Conley, *Irish in Rhode Island*, 13; Hayman, *Catholicism in Rhode Island*, 1:12–18; H. M. Gettleman, "No Irish Need Apply: Patterns of and Responses to Ethnic Discrimination in the Labor Market," *Labor History* 14:1 (winter, 1973); James deBoer, "Paddy in Rhode Island: Perceptions of the Irish in Mid-Nineteenth Century America," unpub. undergrad. paper, History Dept., Brown University, 2005. DeBoer uncovered an amazing, almost daily menu of slurs against the Irish by examining the *PJ* and other newspapers, 117. The appendix includes an index of those articles between 1838 and 1857.

24. Cited in Harvey Strum, "'Not Forgotten in Their Afflictions': Irish Famine Relief from Rhode Island, 1847," *Rhode Island History* 60:1 (winter, 2002), 27; Hayman, MS, ch. 1, p. 20.

25. DeBoer, 38; "Paddy in Rhode Island," Mark S. Schantz, *Piety in Providence: Class Dimensions of Religious Experience in Antebellum Rhode Island* (Ithaca: Cornell University Press, 2000), 244–46.

26. Thomas Man, *A Picture of Woonsocket or The Truth in its Nudity* (n.p., 1835), 32; Thomas Man, *Picture of a Factory Village: To which Are Annexed, Remarks on Lotteries* (Providence: n.p., 1833).

27. *Providence City Directory,* 1838 (Providence: H. H. Brown, 1838), introduction.

28. Schantz, *Piety in Providence,* 169–70.

29. EMG [unidentified] to Dr. Jacob B. White, 3 Nov. 1847, Coll. of Thomas Green; Coleman, *Transformation,* ch. 6. The list of abrasive epithets aimed at Irish Catholics is inexhaustible in that era; these simply provide a sample.

30. Kulik, "Pawtucket Village," 17–31; "Diary of Thomas Jenckes," manuscript coll., Rhode Island Historical Society Library, 25 May 1824; David Zonderman, *Aspirations and Anxieties: New England Workers and the Mechanized Factory System, 1815–1850* (New York: Oxford University Press, 1992); Coleman, *Transformation,* 242.

31. Ed Brown, "Working in Early Rhode Island," a mimeographed set of papers distributed to the delegates at the 1975 AFL-CIO state convention in Providence; Joseph Ott, "Rhode Island Housewrights, Shipwrights, and Related Craftsmen," *Rhode Island History* 31 (1972), 65–79. Editha Hadcock, "Labor Problems in Rhode Island Cotton Mills, 1790–1940," unpub. Ph.D. diss., Brown University, 1945; Teresa Murphy, *Ten Hours' Labor: Religion, Reform and Gender in Early New England* (Ithaca: Cornell University Press, 1992); William Shade, "The Rise of the Providence Association of Mechanics and Manufacturers, A Workingman's Organization, 1799–1850," M.A. thesis, Brown University, 1962. For a longer bibliography from this era, consult Scott Molloy, Eric Barden, Tim McMahon, eds. *A Guide to the Historical Study of Rhode Island Working People* (Kingston: Rhode Island Labor History Society, 1996); Sullivan, "Olney's Lane Riot," 57.

32. *Pawtucket Chronicle,* 13 Sept. 1828; *Rhode Island American and Providence Gazette,* 16 Sept. 1828. David Wilkinson was the brother-in-law of Samuel Slater, Carroll, *Rhode Island,* 4:1150.

33. The phrase "democracy in decline" is the lead title of Patrick T. Conley's *Constitutional History.* Botelho, *Right and Might,* 2:30. Father Hayman believed there were more Irish Catholics employed in textile production at the time, MS, ch. 1, 8–9.

34. Sterne, *Ballots and Bibles,* ch. 1; Conley, *Democracy in Decline,* chs. 9–10; Marvin E. Gettleman and Noel P. Conlon, "Responses to the Rhode Island Workingmen's Reform Agitation of 1833," *Rhode Island History,* 28 (1969), 75–94.

35. Cited in Botelho, *Right and Might,* 1:21.

36. Broadside, "Native American Citizens! Read and Take Warning," reproduced in Botelho, *Right and Might,* 2:33; Scott Molloy, Carl Gersuny, and Robert Macieski, eds., *Peaceably if we can, Forcibly if we must!* (Providence: Rhode Island Labor History Soceiety, 1998), 16–17.

37. Coleman, *Transformation* 220; Gettleman and Conlon, "Responses," 75–94; Hayman, MS, ch. 1, pp. 16–17.

38. Conley, *Democracy in Decline,* ch. 13; Botelho, *Right and Might,* 2, 29–37; Russell J. DeSimone and Daniel C. Schofield, *The Broadsides of the Dorr Rebellion*

(Providence: The Rhode Island Supreme Court Historical Society, 1992), 17–18.

39. Botelho, *Right and Might,* 2:29–37; Conley, *Democracy in Decline,* ch. 13; [Edmund Burke], *Interference of the Executive in the Affairs of Rhode Island* (28 Congress, 1st session, House Report no. 546, serial 447, Washington, 1844), 510. This important document, popularly known as *Burke's Report,* is a treasure trove of primary material; George M. Dennison, *The Dorr War: Republicanism on Trial, 1831–1861* (Lexington: The University Press of Kentucky, 1976), ch. 3.

40. John Brown Francis to Elisha R. Potter Jr., 22 July 1842, Francis Coll., Rhode Island Historical Society, cited in Marvin E. Gettleman, *The Dorr Rebellion: A Study in American Radicalism, 1833–1849* (New York: Random House, 1973), 130.

41. Cited in Botelho, *Right and Might,* 4:70.

42. Joshua Rathbun to Thomas Wilson Dorr, 25 March 1842, cited in Sterne, *Ballots and Bibles,* 30.

43. *Republican Herald,* 12 March 1842, cited in Hayman, MS, ch. 2, 32.

44. Elisha Potter Jr. to John Brown Francis, 1 May 1842, cited in Sterne, *Ballots and Bibles,* 24; Hayman, *Catholicism in Rhode Island* 26, 36–50.

45. Russell DeSimone, "Prisoners of War," unpub. manuscript; *PJ,* 13 July 1842; Sterne, *Ballots and Bibles,* 24. Hayman, MS, ch. 2, 33. Strum, "Not Forgotten," 28; Al Klyberg, "The Impact of the Irish Potato Famine in Rhode Island," guidebook to the play *A State of Hope,* np., n.d.

46. Botelho, *Right and Might,* 2.

47. DeSimone and Schofield, *Broadsides,* 18–19; Botelho, *Right and Might,* 3. Virtually every account of the affair in the twentieth century reflected a liberal interpretation. For the major conservative history of the incident, see Arthur May Mowry, *The Dorr War: The Constitutional Struggle in Rhode Island* (Providence: Preston and Rounds, 1901).

48. Gettleman, *The Dorr Rebellion,* 147; Botelho, *Right and Might,* 3; Conley, *Democracy in Decline,* ch. 13; McLoughlin, *Rhode Island,* 201–205; Matthew J. Smith, "The real McCoy in the Bloodless Revolution in 1935," *Rhode Island History* 32 (1973), 67–85. The Sprague empire would also include a flax mill, the kind that drew some Irish to Rhode Island from Ballybay, Country Monaghan. Lydia L. Rapoza, "Celebrating the Life of William Sprague," *Newsletter,* Cranston Historical Society (September 2007), n.p.

49. The only monograph about the incident is the provocative treatment by Charles Hoffmann and Tess Hoffmann, *Brotherly Love: Murder and the Politics of Prejudice in Nineteenth-Century Rhode Island* (Amherst: University of Massachusetts Press, 1993), 85. A facsimile, second edition of the trial transcript is *The Trial of John Gordon and William Gordon Charged with the Murder of Amasa Sprague* (Providence: Sidney S. Rider, 1884). Testimony provided different months for the family's arrival from Ireland, in June or July 1843, pp. 43, 62, 85. The participants words are recorded in *Republican Herald,*

20 March 1839; McCarron, "Altered States," described these combination groceries and taverns as successful in Rhode Island. *Republican Herald*, 23 March 1839; Hayman, MS, ch. 1, 29.

50. Hoffmanns, *Brotherly Love*, Ch. 1.

51. *Trial of John Gordon*, 77; Hoffmanns, *Brotherly Love*, chs. 1–3. John Brown Francis to Elisha R. Potter Jr., January 9, 14, 1844. Elisha R. Potter Jr. Papers, Rhode Island Historical Society. For an earlier controversial murder involving Irish immigrants in Massachusetts in 1806 see the historical novel by Michael White, *The Garden of Martyrs* (New York: St. Martin's Griffin, 2004).

52. *Trial of John Gordon*, 131. Several wealthy contributors made substantial loans, such as Irish activist Jeremiah Baggott who provided $1,000 and saluted Daniel O'Connell at the first St. Patrick's Day commemoration, *Republican Herald*, 23 March 1839.

53. Cited in Hoffmanns, *Brotherly Love*, 127.

54. *Trial of John Gordon*, 92; Hoffmanns, Chs. 3–4.

55. Hoffmanns, *Brotherly Love*, chs. 3–4, 71; *Trial of John Gordon*, 80.

56. Hoffmanns, *Brotherly Love*, chs. 5–6.

57. Hoffmanns, *Brotherly Love*, ch. 6, 113. Interest in this case remains strong, and several Rhode Island historians held a well-attended seminar about the event in June 2007 at the Sprague Mansion. *Providence Phoenix*, 18 May 2007. Francis to Potter Jr., 14 Jan. 1844, Potter Papers. Father Brady preached at a mass in Providence on St. Patrick's Day in 1845, Hayman, MS, ch. 2, p. 25.

58. Hoffmanns, *Brotherly Love*, ch. 6; Scott Molloy, "Pride, Prejudice and the Gordon Hanging," *PJ*, 17 March 2003.

59. Hoffmanns, *Brotherly Love*, 135; Philip E. Mackey, "'The Results May Be Glorious'; Anti-Gallows Movement in Rhode Island, 1838–1852," *Rhode Island History* 33 (1974), 19–31.

60. Hoffmanns, *Brotherly Love*, 136–40.

61. Hoffmanns, *Brotherly Love*, ch. 8. *Baltimore Republican*, reprinted in the *Republican Compiler*, Gettysburg, Penn., 17 March 1845.

62. Henry Mann, *Our Police: A History of the Providence Force from the First Watchman to the Latest Appointee* (Providence: J. M. Beers, 1889), 230; Hoffmanns, *Brotherly Love*, ch. 8; Molloy, "Brave Irish"; Joseph Sullivan, unpub. manuscript and citations in author's possession; Miller, *Emigrants and Exiles*, introduction.

63. DeSimone and Schofield, *Broadsides*, 76–77.

64. Conley and Smith, *Catholicism in Rhode Island*, 49. Hayman, MS, ch. 2, 38.

65. Hoffmanns, *Brotherly Love*, 162, fn 10; Botelho, *Right and Might*, 4, 70–72.

66. *Trial of John Gordon*, 85.

67. Hoffmanns, *Brotherly Love*, 132; Coleman, *Transformation*, 244; Conley, *The Irish in Rhode Island*, 14. Hayman, MS, ch. 1, 26–27. Carroll, *Rhode Island*, 2:1151. Philip Allen, although a Yankee, hired the Irish to work in his mills.

He also went to the 1839 St. Patrick's Day banquet where another participant thanked him publicly for his "liberality towards the Catholics of Providence, and Irishmen generally." He responded, "Ireland and Irishmen,
may they forever prosper." *Republican Herald,* 23 March 1839.

68. Hoffmanns, *Brotherly Love,* 26; Conley and Smith, *Catholicism in Rhode Island,* 44; Hayman, MS, ch. 2, 69. Andrew Morris, "The Problem of Poverty:
Public Relief and Reform in Postbellum Providence," *Rhode Island History*
45 (1986):1, 1–19.

69. McLoughlin, *Rhode Island,* ch. 1; Hayman, *Catholicism in Rhode Island,* 21–
22; Conley and Smith, *Catholicism in Rhode Island,* 27; Hayman, MS, ch. 1,
25, 81.

70. Cited in Conley and Smith, *Catholicism in Rhode Island* 40–41.

71. Coleman, *Transformation,* 243–45; the author saw Allen as an opportunist.
Conley and Smith, *Catholicism in Rhode Island,* 76–77; McLoughlin, *Rhode Is-
land,* ch. 1; "Allen's Print Works," *PJ,* 17 Dec. 1900; "Death of Hon. Philip
Allen," *PJ,* 18 Dec. 1865, 17 Dec. 1900; William Harold Munro, *Memorial
Encyclopedia of the State of Rhode Island* (New York: The American Historical
Society, 1926), 32.

CHAPTER 3: THE RHODE ISLAND IRISH (pages 59–79)

1. Cited in Harvey Strum, "'Not Forgotten in Their Afflictions': Irish Famine
Relief from Rhode Island, 1847," *Rhode Island History* 60:1 (Winter 2002),
27.

2. Strum, "'Not Forgotten,'" 27; Patrick T. Conley, *Irish in Rhode Island; A Historical Appreciation* (Providence: Rhode Island Heritage Commission and
Rhode Island Publications Society, 1986), 9.

3. Thomas W. Bicknell, *The History of the State of Rhode Island and Providence
Plantations,* 5 vols. (New York: American Historical Society, 1920), 2:1150;
Strum, "'Not Forgotten'" 31–33.

4. Peter J. Coleman, *The Transformation of Rhode Island, 1790–1860* (Providence: Brown University Press, 1969), 146; Bicknell, *Rhode Island,* 5:486;
Edward T. McCarron, "Altered States: Tyrone Migration to Providence
Rhode Island During the Nineteenth Century," *Clogher Record* 16:1 (1997),
149; *Providence Industrial Sites,* n.a. (Providence: Statewide Historical Preservation Report, 1981), 49–50.

5. Philip Foner, "Journal of an Early Labor Organizer," *Labor History* 10
(spring, 1969), 211–12.

6. *Woonsocket Patriot* (hereafter, *WP*), 15 Jan. 1892; Foner, "Journal," 212; Evelyn S. Sterne, *Ballots and Bibles: Ethnic Politics and the Catholic Church in Providence* (Ithaca: Cornell University Press, 2004), ch. 2; Coleman, *Transformation,* 147; Joseph Banigan, "Will," 4663, Providence City Hall.

7. Welcome Arnold Greene, *The Providence Plantations for Two Hundred and
Fifty Years* (Providence: J. A. and R. A. Reid, Publishers and Printers, 1886),
331.

8. Hayman, *Catholicism in Rhode Island*, 118–19, also MS, ch. 2, 40, 65; Patrick T. Conley, *The Irish in Rhode Island: A Historical Appreciation* (Providence: Rhose Island Heritage Commission and Rhode Island Publications Society, 1986), 9, 10, 13, 16; Charles Carroll, *Rhode Island: Three Centuries of Democracy*, 4 vols. (New York: Lewis Historical Publiching, 1932), 4:1151.

9. "Providence Shell Work," *Providence Journal* (hereafter *PJ*), 7 June 1871.

10. *Providence Industrial Sites*, 11–12; Bicknell, *Rhode Island*, 2:486, 3:841–43; *PJ*, 29 July 1898; Coleman, *Transformation*, 150–53; McCarron, "Altered States," 153. The *Providence City Directory* listed Banigan as a jeweler in 1860.

11. Joseph Finnegan placed an ad seeking the whereabouts of his sister-in-law, Ann "Bannagan" of "Carrickadowy" last heard from in Savannah, Georgia, in 1848, in the *Boston Pilot*, a popular Irish-Catholic newspaper with a wide circulation that carried many similar advertisements under "Missing Friends." It appeared on 31 May 1851 and listed Finnegan's address as 16 Orange Street, Providence, the same address as the Banigan family. The *Boston Pilot* is available on microfilm. Ann Banigan is eventually reunited with the family. McCarron, "Altered States," 150–51. Thanks to Anne Duffy for this citation.

12. Carroll, *Rhode Island*, 2:1150; Conley, *Irish in Rhode Island*, 9, 10, 13; Robert A. Wheeler, "Fifth Ward Irish: Immigrant Mobility in Providence, 1850–1870," *Rhode Island History*, 32 (1973), 53–54; McCarron, "Altered States," 153, 157. Dr. Edwin M. Snow, "Statistics and Causes of Asiatic Cholera as It Prevailed in Providence, in the Summer of 1854," *Providence City Documents*, no. 5, 1855, 10–12; John Brown Francis to Elisha S. Potter, 31 Dec. 1857; Francis Papers, 11 Jan. 1855, Series 1, Box 2, Folder 6.

13. Joseph Duffy, ed., "American Journal of James Donnelly," in *Clogher Record Album: A Diocesan History* (Monaghan Town: Clogher Historical Society, 1975), entry for March 1854, 14–15; Hayman, *Catholicism in Rhode Island*, 1, Ch. 3. The article reproducing some of the diary is condensed. On a research trip to County Monaghan in October 2005 I passed the cathedral where the diary is stored but was ignorant of its existence at the time; McCarron, "Altered States," 146.

14. Duffy, "American Journal," 17 March 1854.

15. Duffy, "American Journal," 19, 21 March 1854.

16. Duffy, "American Journal," 4 June 1854.

17. Duffy, "American Journal," 31 March 1854.

18. Unpub. "Diary," Bishop James Donnelly, 27 Oct. 1885, Clogher Diocesan *Archives*, Bishop's House, St. Macartan's Cathedral, County Monaghan, Ireland. My thanks to Peadar Murname, Grace Moloney, and Ann Harney for calling my attention to this entry as well as the article about the bishop's diary. The village of Ballitrain no longer exists but was near Ballybay where the linen bleachers lived who came to Rhode Island.

19. *PJ*, 20 May 1852.

20. *New York Times*, 20 May 1852; *PJ*, 20 May 1852; *Boston Pilot*, 31 May 1851;

Henry Mann, *Our Police: A History of the Providence Police Force from the First Watchman to the Latest Appointee* (Providence: J. M. Beers, 1889), 46–49; Wheeler, "Fifth Ward," 54–55. Thanks to Detective George Pearson of the Providence Police Department for this citation.

21. Livingstone, *The Monaghan Story*, 207; Scott Molloy, "A Long Struggle to Enter the Mainstream," *PJ*, 17 March 1997; Patrick T. Conley and Paul R. Campbell, *Firefighters and Fires in Providence: A Pictorial History of the Providence Fire Department, 1754–1984* (Norfolk, Va.: Providence Publications Society, 1985), 25–31; Wheeler, "Fifth Ward," 54; Hayman, MS, ch. 2:54–56, 63; *Providence Post*, 10 May 1859.

22. John R. Mulkern, *The Know-Nothing Party in Massachusetts: The Rise and Fall of a People's Movement* (Boston: Northeastern University Press, 1990); Nancy Lusignan Schultz, *Fire and Roses: The Burning of the Charlestown Convent, 1834* (New York: The Free Press, 2000); Patrick T. Conley, "The Irish of Rhode Island," in Patrick T. Conley, *Rhode Island in Rhetoric and Reflection: Public Addresses and Essays* (East Providence: Rhode Island Publications Society, 2002), 26; Thomas H. O'Connor, *The Boston Irish: A Political History* (Boston: Northeastern University Press, 1995), ch. 3; Hayman, *Catholicism in Rhode Island* 1:135–40. For a history of the Sisters of Mercy in the state, see *Seventy Years in the Passing with the Sisters of Mercy, Providence, Rhode Island, 1851–1926*, n.a. (Providence: Providence Visitor Press, 1926), 85–87; Michael A. Simoncelli, "Battling the Enemies of Liberty: The Rise and Fall of the Rhode Island Know-Nothing Party," *Rhode Island History* 54 (1996), 3–19; John Michael Ray, "Know Nothingism in Rhode Island," *American Ecclesiastical Review* 148 (1963), 27–38.

23. Duffy, "American Journal," 14 Nov. 1854; McCarron, "Altered States," 145; *Providence Post*, 21 May 1859.

24. Wheeler, "Fifth Ward," 56; Bicknell, *Rhode Island*, 2, 842.

25. Joseph Banigan and George W. Miller, "Improvement in Moulds for Vulcanizing Rubber Pencil-Tips," United States Patent Office, No. 99,045, 18 Jan. 1870; Henry C. Pearson, "The India-Rubber Industry in New England," in William T. Davis, ed., *The New England States*, 4 vols. (Boston: D. H. Hurd and Co., 1897), 1:334–53.

26. Pearson, "India-Rubber," 1:334.

27. Pearson, "India-Rubber," 1:339–42; Charles Goodyear, "Gum-Elastic and Its Varieties, With a Detailed Account of Its Applications and Uses and of the Discovery of Vulcanization," reprinted in *A Centennial Volume of the Writings of Charles Goodyear and Thomas Hancock* (American Chemical Society, 1939).

28. Pearson, "India-Rubber," 1:340; Richard Korman, *The Goodyear Story: An Inventor's Obsession and the Struggle for A Rubber Monopoly* (San Francisco: Encounter Books, 2002), 22–23, 69–71, 169–70. A Providence native, Stephen G. C. Smith, actually received the first patent for making rubber footwear in 1837, *The Story of Rubber*, n.a. (Boston: Hood Rubber Company, ca. World War I), 5.

29. "A Veteran Rubber Manufacturer," *Boot and Shoe Reporter* (hereafter *BSR*), 23 June 1897. Surprisingly, Peter Coleman missed the importance of the rubber industry locally. *The Biographical Encyclopedia of Representative Men of Rhode Island* (Providence: National Biographical Publishing, 1881), 354–55. *Biographical Cyclopedia,* biography of Hartshorne, 337–38; *Biographical Cyclopedia,* biography of Bourn, 353–54; *BSR,* 7 July 1897; Korman, *Goodyear Story* 185–86, 188; *Goodyear vs. The Providence Rubber Company;* Abraham Payne, *Reminiscences of the Rhode Island Bar* (Providence: Ticknor and Preston, 1885), ch. 27, 30; the author explored the rubber cases in which he served as counsel. Ironically, Thomas Carpenter, former Democratic nominee for Governor during the Dorr War era as well as counsel to Dorr and the Gordons, served as Payne's mentor, 223. Ironic, because George Bourne served as a first lieutenant in the state militia as an avid Law and Order advocate as did Dr. Isaac Hartshorne, proprietor in another rubber lawsuit defended by Payne.

30. Exactly when Banigan worked at the Bourn Company is unknown. He probably toiled there around 1860 when he married Margaret Holt, but the *Woonsocket Reporter* claimed the company employed him in 1867 after his stint with Haskins, which does not jibe with the offer from Woonsocket that lured him from the Massachusetts enterprise, *Woonsocket Reporter* (hereafter cited as *WR*), 29 July 1898.

31. "Providence Rubbers," *BSR,* 28 July 1897; Richard M. Bayles, ed., *History of Providence County, Rhode Island,* 2 vols. (New York: W.W. Preston, 1891), 1:401; *Providence Visitor* (hereafter *PV*), 22 Feb. 1896; *PJ,* 17 Feb., 13, 14 March 1896; Bayles' biographical sketch is almost identical to the obituary in the *PJ;* he may have written the death notice earlier; *Record of Proceedings before Charles Hart, Esq., Master, in the Case of Charles Goodyear, Providence Rubber Company, et. al.* (Pawtucket: R. Sherman and Co., 1866), 261–74, 303; Pearson, "India-Rubber," 338.

32. Marriage Records, City of Providence; Wheeler, "Fifth Ward," 57–58 (citing Oscar Handlin); *Providence City Directories,* 1859, 1866, 1868; Hayman, *Catholicism in Rhode Island* 1,104–105; James F. Reilly, *Souvenir Booklet Commemorating the One Hundredth Anniversary of St. Joseph's Parish* (Providence: privately printed, 1951).

33. *India Rubber World* (hereafter *IRW*), 1 Aug. 1898.

34. *Woonsocket Call,* 3 Feb. 1975; Charles Slack, *Noble Obsession: Charles Goodyear, Thomas Hancock, and the Race to Unleash the Greatest Industrial Secret of the Nineteenth-Century* (New York: Hyperon Books, 2002), ch. 3; *PJ* 29 July 1898; *New York Times,* 29 July 1898.

35. *IRW,* 1 Aug. 1898.

36. *PJ,* 29 July 1898; Dr. Alton B. Thomas, an amateur historian, wrote about Banigan in the 1970s. The physician had conversations and probably access to documents owned by Otto Koerner, an industrial relations manager with the United States Rubber Company in Woonsocket, successor to the

Banigan complex there, *Woonsocket Call,* 3 Feb. 1975; *New York Times,* 29 July 1898; Pearson, "India-Rubber," 343; Slack, *Noble Obsession,* ch. 3.

37. "Birth Certificates," Boston and Roxbury, Mass.; Woonsocket, R.I.; Obituary, Alice Banigan Sullivan, *PJ,* 29 Sept. 1909.

38. Kerby Miller, *Emigrants and Exiles: Ireland and the Irish Exodus to North America* (New York: Oxford University Press, 1985), 359–61.

39. Patrick T. Conley and Matthew J. Smith, *Catholicism in Rhode Island,* 91; Don Skuce, "Rhode Island's Irish Regiment," lecture and handouts, Warwick Historical Society, 24 April 1997; Hayman, *Catholicism in Rhode Island,* 1:150–54.

40. Hayman, *Catholicism in Rhode Island: The Formative Era* (Providence: Diocese of Providence, 1976), 1:154; Conley, *Irish in Rhode Island,* 15.

41. *PJ,* 25 Aug. 1864; see *Providence Manufacturers and Farmers Journal,* for Black and Irish quarrels just in 1865, 12 Jan., 13, 17, July, 10, 14 Aug., 2, 30 Oct., 24 Nov., 18 Dec. 1865. Conley and Smith, *Catholicism in Rhode Island,* 89–93; Conley, *Irish in Rhode Island;* Billhead, Emmet Guards, in author's possession; Hayman, *Catholicism in Rhode Island,* 1:154–58; Bicknell, *Rhode Island,* 2:622–23; Frederick Denison, *Shot and Shell: The Third Rhode Island Heavy Artillery Regiment in the Rebellion, 1861–1865* (Providence: J. A. & R. A. Reid, 1879); Sterne, *Ballots and Bibles,* 61–62; Carroll, *Rhode Island,* 2:1150; P. A. Sinnott, "Soldiers' Franchise," the flyer has no publisher or date and is part of the private Russell DeSimone Rhode Island Coll.; *WP,* 22 Feb. 1878, 9 Dec. 1882, 5 April 1893, 9 Aug. 1895.

42. Patrick T. Conley, "Rhode Island and the Emasculation of the Fifteenth Amendment," *Rhode Island Bar Journal* (September–October, 2007), 25–27.

43. Albert K. Aubin, *The French in Rhode Island* (Providence: The Rhode Island Heritage Commission, 1988); Patrick T. Conley, *An Album of Rhode Island History, 1636–1986* (Norfolk, Virginia: The Donning Company/Publishers, 1986), ch. 5. Raymond Sickinger and John Primeau, *The Germans in Rhode Island: Pride and Perseverance, 1850–1895* (Providence: The Rhode Island Heritage Commission, 1985).

44. Aubin, *French in Rhode Island,* 11–16.

45. Wheeler, "Fifth Ward," 54; Cottrol, *Rhode Island,* 114; *PJ,* 30 Nov. 1867; *WP,* 23 Aug. 1867; Edwin M. Snow, M.D., *Report Upon the Census of Rhode Island, 1875* (Providence: Providence Press Company, 1877), xliii.

46. *Providence Morning Star* (hereafter *PMS*), 4 March 1879, 6 Dec. 1880, 5 Jan. 1881; *PJ,* 31 July, 11 Nov. 1867, 29 Feb., 27 July 1868; Conley, *Democracy in Decline;* Fionnghuala Sweeney, "'The Republic of Letters,' Frederick Douglas, Ireland, and the Irish Narratives," in Kevin Kenney, ed., *New Directions in Irish-American History* (Madison: University of Wisconsin Press, 2003), 123–39; L. Perry Curtis Jr., *Apes and Angels: The Irishman in Victorian Caricature* (Washington, D.C.: Smithsonian Institution Press, 1997); Joseph Sullivan, *Marxists, Militants, and Macaroni: The I.W.W. in Providence's Little Italy* (Kingston: The Rhode Island Labor History Society, 2000); Douglas

quoted in Brian Dooley, "Black and Green: The Historic Connection between Black American Civil Rights and Irish Nationalists," *Irish America Magazine* (Sept./Oct. 1998), 32; see Dooley's book for greater detail, *The Fight for Civil Rights in Northern Ireland and Black America* (Chicago: Pluto Press, 1998).

47. W. E. B. DuBois, *Dusk of Dawn* (New Brunswick, N.J.: Transaction Publishers, 1984), 14; Noel Ignatiev, *How the Irish Became White* (New York, Routledge, 1995), 311–12.

48. *Providence Post,* 7 May, 27 Dec. 1860, 4 April, 19 March, 13 May 1861; Conley, *Album of Rhode Island,* 117–18; Gary Gerstle, *Working-Class Americanism: The Politics of Labor in a Textile City, 1914–1960* (New York: Cambridge University Press, 1989), introduction, ch. 1; Alan M. Kraut, *The Huddled Masses: The Immigrant in American Society, 1880–1921* (Arlington Heights, Il: Harlan Davidson, Inc., 1982).

49. An impressive look at the rise of the Irish-American workforce can be discerned in the number of Irish names associated with the labor movement in Rhode Island in the Gilded Age. See the unattributed compilations: *First Grand Reunion and Ball of the Rhode Island Protective Association under the Auspices of District Assembly 99, K[nights] of L[abor]* (Providence, 1893); *The Illustrated History of the Rhode Island Central Trades and Labor Union and Affiliated Unions* (Providence, 1899); *20th Century Illustrated History of Rhode Island and the Rhode Island Central Trades and Labor Union and Its Affiliated Organizations* (Providence, 1901). An older, Marxist analysis of labor development in the era, Philip S. Foner, *History of the Labor Movement in the United States: From Colonial Times to the Founding of the American Federation of Labor* (New York: International Publishers, 1947); an updated account is American Social History Project, *Who Built America?* (New York: Worth Publishers, 2000), 2 vols.

CHAPTER 4: THE WOONSOCKET RUBBER COMPANY
(pages 80–105)

1. *Woonsocket Patriot* (hereafter *WP*), 9 Aug. 1867; Charles W. Smythe, *History of the Catholic Church in Woonsocket and Vicinity from the Celebration of the First Mass in 1828, to the Present Time* (Woonsocket: Charles E. Cook, Printer, 1903), 82–84.

2. *WP,* 6 Aug. 1866, 21 April, 1 Nov. 1867. Banigan and Cook may have forced Bailey from his wringing company, *WP,* 6 Feb. 1880.

3. John Holt, United States Patent Office, No. 49,030, 25 July 1865.

4. Lyman Cook, sketch in *Biographical Encyclopedia of Representative Men of Rhode Island* (Providence: National Biographical Publishing, 1881), 285–87; *India Rubber World,* (hereafter *IRW*), 1 Aug. 1893; Richard M. Bayles, ed., *History of Providence County, Rhode Island,* 2 vols. (New York: W.W. Preston, 1891), 1:682–84, 2:321–24; *WP,* 20 Dec. 1895.

5. Dun, *Report,* 15 Feb. 1882.
6. *WP,* 31 May 1867, 29 April 1868, 7 Jan. 1887; *Petition for Naturalization,* Rhode Island Supreme Court, Judicial Records Center, Pawtucket, R.I., March term, 1867; Glenn D. Babcock, *History of the Untied States Rubber Company: A Case Study in Corporation Management* (Bloomington: Graduate School of Business, Indiana University, 1966), 435.
7. *IRW,* 1 Aug. 1898.
8. Joseph Banigan, United States Patent Office, No. 97266, 30 Nov. 1869; No. 250871, 13 Dec. 1881; No. 298546, 13 May 1884.
9. Charles Carroll, *Rhode Island: Three Centuries of a Democracy,* 4 vols. (New York: Lewis Historical Publishing, 1932), 2:879.
10. *Encyclopedia,* 286–87; Thomas W. Bicknell, *The History of the State of Rhode Island and Providence Plantations,* 5 vols. (New York: American Historical Society, 1920), 5:485–88; Carroll, *Rhode Island,* 2:879; *WP,* 29 April, 29 May, 25 Dec. 1868, 26 March 1875; Robert Grieve, *The Cotton Centennial, 1790–1890* (Providence: J.A. and R.A. Reid, 1891), 140.
11. *WP,* 15 June 1866, 1, 8 Feb. 1867, 13 Dec. 1872, 18 April 1873, 25 June 1875.
12. *WP,* 7 June 1867; Edwin M. Snow, ed., *Report upon the Census of Rhode Island, 1875* (Providence: Providence Press Company, 1877), lxiv. On occasion, different towns around the state "dumped" heavily populated areas inhabited by Irish Americans into adjoining cities through the process of annexation, Patrick T. Conley and Paul Campbell, *Providence: A Pictorial History* (Norfolk, Va.: The Donning Company, Publishers, 1982), 98–99.
13. *WP,* 24 Jan., 20 Nov. 1868, 10, 24 March, 26 May, 18 Aug. 1871.
14. *WP,* 9, 16 July1869, 15 Feb. 1870, 19 June 1874, 15 Feb. 1878; *Woonsocket, Rhode Island,* n.a. (Providence: Statewide Historic Preservation Report, 1976); Ray Bacon, "The New City, 1888–1900," in Woonsocket Centennial Committee, *Woonsocket, Rhode Island: A Centennial History, 1888–1988* (State College, Penn.: Jostens Printing, 1988), 11–31.
15. Donald Demers, "Edward Harris: Cassimere, Libraries, and the Free Soil Party," Providence College, M.A. thesis, 2003. Harvey Strum, "'Not Forgotten in Their Afflictions': Irish Famine Relief from Rhode Island, 1847," *Rhode Island History* 60:1 (winter 2002), 31.
16. *WP,* 8 May 1874, 2 June, 10 Nov. 1876; N. A., *The Constitution of Rhode Island and Equal Rights,* n.a. (Providence: Equal Rights Association, 1881). Gary Gerstle, *Working-Class Americanism: The Politics of Labor in a Textile City, 1914–1960* (Princeton: Princeton University Press, 2002).
17. *WP,* 24 March 1871, 8 Aug. 1873, 20, 26 Nov. 1874, 19 March, 10 Dec. 1875, 15 July, 7 Oct. 1881.
18. Al Klyberg, "The Impact of the Irish Potato Famine in Rhode Island," guidebook to the play *A State of Hope,* n.p., n.d.
19. Richard M. Bayles, ed., *History of Providence County, Rhode Island,* 2 vols. (New York: W. W. Preston, 1891), 2:401.

20. Bayles, *History of Providence County*, 2:401; N.A., Works Progress Administration, *Rhode Island: A Guide to the Smallest State* (New York: Houghton Mifflin Co., 1937), 313; *Record of Proceedings Before Charles Hart*, 262; *Providence Evening Bulletin (hereafter PEB)*, 13 and 14 March 1896; *Providence Visitor* (hereafter *PV*), 22 Feb. 1896; *Providence Journal* (hereafter *PJ*), 17 Feb. 1896.

21. *WP*, 1 Oct., 17 Dec. 1870, 12 May 1871, 8 Dec. 1876; *Dun Reports*, Providence County, vol. 6, 11 Sept. 1873, Baker Library, Harvard University.

22. *WP*, 9 Aug. 1878.

23. *WP*, 20 June 1873, 11 Dec 1874, 30 April, 30 July, 1875, 1 April 1881, 12 Oct. 1883; William G. McLoughlin, *Rhode Island: A Bicentennial History* (New York: Norton, 1978), 166–68; Peter J. Coleman, "The Entrepreneurial Spirit in Rhode Island History," *Business History Review* 37:4 (winter, 1963), 319–44.

24. Carl Gersuny, "A Unionless General Strike: The Rhode Island Ten-Hour Movement of 1873," *Rhode Island History* 54 (1996), 21–32.

25. *PJ*, 30 May 1873, cited in Gersuny, "A Unionless Strike," 30; There are many references to the role of English immigrants from Lancashire in the regional labor movement, Mary Blewett, *Constant Turmoil: The Politics of Industrial Life in Nineteenth-Century New England* (Amherst: University of Massachusetts Press, 2000).

26. Dun, *Report*, vol. 6, 20 Sept. 1876, spelling, punctuation, and abbreviations as in the original; *WP*, 31 Oct., 21 Nov., 5 Dec. 1873, 26 Feb., 12, 19 March, 15, 21 May, 1, 29 Oct. 1875, 21 April 1876.

27. *WP*, 29 Oct. 1875, 11 Feb., 7, 21 April, 27 Oct. 1876, 12, 19, Jan. 1877, 1 April 1881.

28. *WP*, 9 Dec. 1881, Dun, *Report*, 20 Sept. 1876.

29. *WP*, 2, 16 Feb., 20 April, 8 June, 31 Aug., 12 Oct., 16 Nov., 14 Dec. 1877.

30. *WP*, 20 April 1877.

31. *WP*, 20 April 1877.

32. *WP*, 20 April 1877, 6 Feb. 1885; Daniel Nelson, *American Rubber Workers and Organized Labor 1900–1941* (Princeton: Princeton University Press, 1988), ch. 2; B. G. Underwood, "Manufacture of Rubber Shoes," *Scientific American* 10 (1892), 374–75.

33. Dun *Report*, vol. 6, 14 March 1877, 4 May 1880.

34. *WP*, 1 Nov. 1878, 20 June, 22 Aug., 5 Dec. 1879, 12 Aug. 1881, 24 Feb., 15 Dec. 1882, 13 April, 11 May, 16 March 1883, 11 April 1884.

35. *Providence Post*, 11 Sept. 1857. Thanks to Reverend Robert Hayman for this citation. *WP*, 20 June 1879, 23 Jan. 1880, 3 Nov. 1882, 13 April, 11 May 1883, 11 April 1884; *Woonsocket Evening Reporter* (hereafter *WER*), 20 Dec. 1885.

36. Dun, *Report*, 15 Feb. 1882; *WP*, 17 Feb. 1882, 10 July 1896.

37. *WP*, 3 March 1882, 22 Feb., 14 March, 9 May, 3 Oct. 1884.

38. Robert Evans and Samuel Colt, United States Rubber Company, "Report of the Special Committee," 1893, Baker Library, Harvard University; *Boot and*

Shoe Recorder (hereafter *BSR*), 1 Feb. 1893; *WP,* 22 July, 16 Dec. 1881, 1 Sept. 1882, 16 March 1883.

39. American Social History Project, *Who Built America?,* 2 vols. (New York: Worth Publishers, 2000).

40. Maury Klein, *The Life and Times of Jay Gould* (Baltimore: Johns Hopkins Press, 1986), 357–63; Paul Buhle, "The Knights of Labor in Rhode Island," *Radical History Review* 17 (spring, 1978), 75–98.

41. Terence Powderly, *The Path I Trod: The Autobiography of Terence Powderly* (New York: Columbia University Press, 1940), ch. 11; Robert E. Weir, "When Friends Fall Out: Charles Litchman and the role of Personality in the Knights of Labor," in Kenneth Fones-Wolf and Martin Kaufman, eds. *Labor in Massachusetts: Selected Essays* (Westfield, Mass.: Westfield State College, 1990).

42. *Report of the Fourth Annual Session: Federation of Trades and Labor Unions of the United States and Canada* (Washington, D.C.: Gray and Clarkson, Printers, 1884), 16–21; Tom Juravich, William F. Hartford, and James R. Green, *Commonwealth of Toil: Chapters in the History of Massachusetts Workers and Their Unions* (Amherst: University of Massachusetts Press, 1996), chs. 6–7. As early as 1869 an Irish Laborers Union marched in Providence's Fourth of July Parade as well as the St. Patrick's Day procession, *Manufacturers and Farmers Journal,* 8 July 1869, 22 Aug. 1870; *PJ,* 18 March 1871. Anthony Lementowicz, "Trouble in Labor's Eden: Labor Conflict in the Quarries of Westerly, 1871–1922," *Rhode Island History* 59:1, (Feb. 2001) 19–30; Craig Phelan, *Grand Master Workman: Terence Powderly and the Knights of Labor* (Westport, Conn.: Greenwood Press, 2000), 3. Bruce Laurie, *Artisans into Workers: Labor in Nineteenth-Century America* (Urbana, Ill.: University of Illinois Press, 1997), 157. Robert Weir, *Beyond Labor's Veil: The Culture of the Knights of Labor* (University Park, Penn.: Pennsylvania State Press, 1996), ch. 1 esp. "The [rubber] industry is one which, though just across the border in Massachusetts, is eminently a Rhode Island industry." *Providence Morning Star* (hereafter *PMS*) 21 Sept. 1885. Class antagonisms began with the birth of the industrial revolution in Rhode Island, Gary Kulik, "Pawtucket Village and the Strike of 1824: The Origins of Class Conflict in Rhode Island," *Radical History Review* 17 (spring 1978), 5–37; Paul Buhle, "The Knights of Labor in Rhode Island" *Radical History Review* 17 (1978); George E. McNeill, *The Labor Movement: The Problem of Today* (Boston: A. M. Bridgeman, 1887); Stuart B. Kaufman, ed., *The Samuel Gompers Papers: The Making of a Union Leader, 1850–1886* (Chicago: University of Illinois Press, 1986); Nick Salvatore, ed., Samuel Gompers, *Seventy Years of Life and Labor: An Autobiography* (Ithaca: ILR Press, 1984), the editor's introduction is particularly insightful. Alexander Keyssar, *The Right to Vote: The Contested History of Democracy in the United States* (New York: Basic Books, 2000), 29, 130. Lincoln Steffens exposed the system: "Rhode Island: A State for Sale," *McClure's Magazine* 24 (Feb. 1905); Norman J. Ware, *The Labor Movement in the*

United States, 1860–1895 (New York: D. Appleton and Company, 1929), 97–102; Patrick T. Conley, *The Irish in Rhode Island: A Historical Appreciation* (Providence: Rhode Island Heritage Commission and Rhode Island Publications Society, 1986), 16.

43. *WER*, 26 Oct. 1885; *PV*, 30 July 1898. *Boston Citizen*, 3 March 1894. *Rhode Island State Census, 1885* (Providence, 1887), 95, 246, 353, 356–57. In an interview with John Parker, 10 May 2006 he mentioned that Brown and Sharpe kept a "No Irish Need Apply" sign outside the employment office on Promenade Street in Providence early in the twentieth century. For another look at this ethnic working class in the same time period, see Patricia Kelleher, "Young Irish Workers: Class Implications of Men's and Women's Experiences in Gilded Age Chicago," in Kevin Kenny, ed., *New Directions in Irish-American History* (Madison: University of Wisconsin Press, 2003).

44. Cardinal Simeoni announced Banigan's selection in a letter to Bishop Thomas Hendricken, 3 May 1885, Diocese of Providence, *Archives*. Simeoni later granted "conditional tolerance" to the Knights in the organization's twilight. Henry J. Browne, *The Catholic Church and the Knights of Labor* (Washington, D.C.: The Catholic University of America, 1949), 26, 108, 324. Robert Weir, *Beyond Labor's Veil: The Culture of the Knights of Labor* (University Park: Pennsylvania State Press, 1996), 94–96. *WP* 8, 22 Sept. 1882.

45. *PJ*, 10, 26 Jan., 3, 6, 16, 27 Feb., 14 April, 20 Aug., 6 Oct., 18 Dec. 1885. *The People*, 25 Dec. 1886. *WER*, 13, 14 Feb., 14 March, 17 Aug. 1885. *WP*, 24 Dec. 1880, 25 Feb., 16 Dec. 1881. *PMS*, 24 Aug. 1881. For the importance of Irish-American workers in union organizing, see Robert Sean Wilentz, "Industrializing America and the Irish: Towards the New Departure," *Labor History* 20:9 (fall, 1979), 585. The Rhode Island Irish were prominent contributors to Irish causes, Eric Foner, "Class, Ethnicity, and Radicalism in the Gilded Age: The Land League and Irish America," *Marxist Perspectives* 1:2 (summer, 1978), 21, 23. Conley, *Irish in Rhode Island*, 15. For the Irish in nearby Worcester, Mass., see Timothy Meagher, *Inventing Irish America: Generation, Class, and Ethnic Identity in a New England City* (Notre Dame, Ind: University of Notre Dame Press, 2001). In the case of Anaconda Copper Company in Butte, Irish management employed ethnic connections in a positive fashion in order to build bridges to their working-class countrymen, David Emmons, *The Butte Irish: Class and Ethnicity in an American Mining Town, 1875–1925* (Urbana, Ill.: University of Illinois Press, 1989).

46. Conley and Smith, *Catholicism in Rhode Island*, 94–96. "The Catholic Church in Rhode Island was de facto an Irish national church," 112. Kim Voss, *The Making of American Exceptionalism: The Knights of Labor and Class Formation in the Nineteenth Century* (New York: Cornell University Press, 1993). The author cited community organizing as being as effective as shop floor activity for the Knights, 4, 142. Scott Molloy, *Trolley Wars: Streetcar Workers on the Line* (Washington: Smithsonian Institution Press, 1996), see ch. 2 for the social role of streetcar workers in Rhode Island neighborhoods. Leon Fink inves-

tigated political action by the Knights in several different states and pointed out the influence of Irish Americans, *Workingmen's Democracy: The Knights of Labor and American Politics* (Urbana, Ill.: University of Illinois Press, 1983), 52, 214.

47. *WP,* 29 Oct. 1885. Joe Doherty, *Woonsocket Call,* 4 Jan. 1998. Doherty wrote a series of weekly retrospective articles about the strike and other Blackstone Valley history using newspaper sources almost exclusively, and conveniently reprinting some articles entirely. Banigan's wealth and influence, although unusual in Gilded Age America, was not unique. Consult the story of John Roche and his efforts among the Irish poor in Iver Bernstein, *The New York City Draft Riots of 1863: Their Significance for American Society and Politics in the Age of the Civil War* (New York: Oxford University Press, 1990).

48. *WP,* 23 Jan. 1880; *The Woonsocket Directory,* 1880–1881 (Providence: E. S. Metcalf, 1880). Relatives and friends often boarded in these homes. Banigan bragged that his help made so much money they could afford their own places, *PJ,* 1 July, 20 Dec. 1885; *WER,* 1 Sept. 1885; *PMS,* 30 Dec. 1880; *WP,* 23 Jan., 18 Dec. 1885, 1 Oct. 1886. A general article about working-class home ownership in Woonsocket is in *PJ,* 24 Nov. 1889; *Rhode Island Census, 1885,* (Providence: E.L. Freeman and Son, 1887), 93, 07, 363.

49. *PJ,* 1 Jan. 1885; *PV,* 3 Jan. 1885; *WP,* 4 April 1884. Bishop Thomas Hendricken to John O'Keefe, published in *The People,* Supplement, 13 Feb. 1886. The Knights responded with a kind obituary of the Bishop as a friend of workers, *The People,* 19 June 1886, 3 Jan. 1885.

CHAPTER 5: A KNIGHT OF ST. GREGORY AGAINST THE KNIGHTS OF LABOR (pages 106–39)

1. *Providence Morning Star,* (hereafter *PMS*), 4, 6 Feb. 1885; *Woonsocket Evening Reporter* (hereafter *WER*), 10 July 1885.

2. *WER,* 2 Feb. 1885; *Providence Journal* (hereafter *PJ*), 3, 6, 14 Feb., 5 May 1885; *PMS,* 3 Feb. 1885.

3. *WER,* 10 July 1885; *PMS,* 6 Feb. 1885; *Woonsocket Call* (hereafter *WC*), 1 March 1998; *Boot and Shoe Reporter* (hereafter *BSR*), 22 March 1893, 127.

4. *PMS,* 3, 6 Feb., 15 July 1885; *WER,* 13 Feb. 1885. Letter to the editor from Charles W. Hoadley, *WER,* 18 July 1885. *PJ,* 4 Feb., 1 July, 1 Oct. 1885. Harold S. Roberts incorrectly identified a New Jersey assembly as the first rubber local in the Knights of Labor, *The Rubber Workers: Labor Organization and Collective Bargaining in the Rubber Industry* (New York: Harper and Brothers, 1944), 24. Striking leather workers in New Jersey in 1887 also finished their work to avoid spoilage and negative publicity, Kim Voss, *The Making of American Exceptionalism: The Knights of Labor and Class Formation in the Nineteenth Century* (Ithaca: Cornell University Press, 1993), 215. *Rhode Island State Census, 1885* (Providence: E.L. Freeman and Son, 1887) cited the average wages for male rubber workers under age 15 at 67 cents a day; over 15, $1.58. Women earned $1.12 over the age of 15; 596. "The strikes in

1885 were far more spontaneous and unorganized than was true of earlier outbreaks," Henry J. Browne, *The Catholic Church and the Knights of Labor* (Washington, D.C.: The Catholic University of America, 1949), 120. In rubber, seventeen of twenty walkouts before 1886 were not ordered by a union. *Third Annual Report of the Commissioner of Labor, 1887* (Washington, D.C.: Government Printing Office, 1888) 338, 814.

5. *Woonsocket Patriot* (hereafter *WP*), 9 Feb., 18 April 1883; *PJ*, 3 Feb. 1885; *WER*, 17 Feb. 1885. One article claimed he never cut wages. *WP*, 21 Feb. 1896; Richard M. Bayles, eds., *History of Providence County, Rhode Island*, 2 vols. (New York: W.W. Preston, 1891), 2:401; Joe Doherty, *WC*, 11 Jan., 15 Feb. 1998; James W. Smyth, *History of the Catholic Church in Woonsocket and Vicinity from the Celebration of the First Mass in 1828, to the Present Time* (Woonsocket: Charles E. Cook, Printer, 1903), 71.

6. *PJ*, 3, 6, 14, 19 Feb. 1885; *PMS*, 18 Feb. 1885; *WER*, 13 Feb. 1885; *Massachusetts Bureau of Labor Statistics: Nineteenth Annual Report, 1888* (Boston: Wright and Patter, 1888), 26–27; *WER*, 26, 27 Aug. 1885.

7. *PMS*, 15 July 1885; *WER*, 13, 17 Feb., 29 June, 10 July, 26, 27 Aug., 2, 17 Sept. 1885; *PJ*, 1 July 1885; Joe Doherty, *WC*, 4 Jan, 8 Feb., 22, 29 March 1998. Banigan was often in court over disputes with suppliers, *Samuel Colt Papers*, Folder 41, Special Collections, University of Rhode Island. A short parody of a Shakespeare play in *India Rubber World* (hereafter *IRW*) referred to the Earl of Woonsocket as "a Hustler," 15 July 1890, 224. Henry C. Pearson, *The Rubber Country of the Amazon* (New York: The India Rubber World, 1911), 58, 217; *PJ*, 3, 20, 21 Feb. 1885; *Worcester Sunday Telegram*, 20 Sept. 1885; *Boston Pilot*, 28 Feb. 1885.

8. *PJ*, 4, 9 May 1885; *WER*, 7, 22, 24 April, 12, 13, 20 June, 7, 10 July 1885; *PMS*, 9, 13 May 1885. *The People*, 29 May 1886.

9. *PJ*, 5 May, 1 July, 22 Aug. 1885; *WER*, 29, 30 June, 8, 9 July 1885; *PMS*, 19 Sept. 1885. Jonathan Garloch, *Guide to the Local Assemblies of the Knights of Labor* (Westport, Conn.: Greenwood Press, 1982), 206, 478. Thomas Cafferty, who led the February walkout, tried to flex the new union's muscle by convincing the Providence and Worcester Railroad to reduce the 15 cents fare for the 4 mile roundtrip between Woonsocket and Millville, *WER*, 13 May 1885. The union, almost from the beginning, was tackling outside issues related to the workplace. The *IRW* cited a similar case of dismissal of a labor activist for poor work in a New Haven, Conn., rubber company, 15 April 1892. *Third Annual Report of the United States Commissioner of Labor, 1887* (Washington, D.C.: Government Printing Office, 1888), 814.

10. Kim Voss, *The Making of American Exceptionalism: The Knights of Labor and Class Formation in the Nineteenth Century* (New York: Cornell University Press, 1993), ch. 3. For other interpretations of the Knights see especially: Robert Weir *Beyond Labor's Veil: The Culture of the Knights of Labor* (University Park: Pennsylvania State Press, 1996); Leon Fink, *Workingmen's Democracy: The Knights of Labor and American Politics* (Urbana: University of Illinois Press,

1983); George S. Kealey and Bryan D. Palmer, *Dreaming of What Might Be: The Knights of Labor in Ontario, 1880–1890* (Cambridge, Mass.: Cambridge University Press, 1982).

11. *WP,* 18 March 1887. Tom Juravich, William F. Hartford, and James R. Green, eds., *Commonwealth of Toil: Chapters in the History of Massachusetts Workers and Their Unions* (Amherst: University of Massachusetts, 1996); Paul Buhle, "The Knights of Labor in Rhode Island," *Radical History Review* 17 (1978), 40; Bruce Cohen, "Lords of Capital and Knights of Labor: Worcester's Labor History During the Gilded Age," *Historical Journal of Massachusetts* (winter, 2001).

12. *WER,* 30 June 1885.

13. *PJ,* 1, 2, 3 July 1885; *WER,* 1, 2, 3 July, 14 Sept. 1885. Banigan would claim that Hines was one of nine employees who worked on the same stock but the only one whose product was deficient. *WER,* 19 Sept. 1885; *PV,* 14 Feb. 1885. The pastor at St. Augustine's was very supportive of the strikers. The records of the church were lost in a fire. *BSR,* 22 March 1893, 127; *The People,* 29 May 1886; David Nelson, *American Rubber Workers and Organized Labor, 1900–1941* (Princeton: Princeton University Press, 1988), 10.

14. *John Swinton's Paper,* 15 Feb. 1885. A shillelagh, a stylized Irish club or walking stick, presented to a foreman was stolen soon after, *WER,* 2 June 1885. Factory "girls" at the American Rubber Company in Cambridge struck in August over an abusive female superintendent, *WER,* 10 Aug. 1885. Joseph Hines recalled being very cautious in his work as he knew he was a marked man, *The People,* 29 May 1886.

15. Henry C. Pearson, "The India-Rubber Industry in New England," in *The New England States: Their Constitutional, Judicial, Educational, Commercial, Professional, and Industrial History,* 4 vols. (Boston: D.H. Hurd, 1897), 1:334–53; *PJ,* 3 July 1885; *WP,* 31 March 1882, 2 Feb., 27 July 1883, 12, 26 July 1889, 1 April 1892, 11 Aug. 1893; *WER,* 22 April 1885; *Framingham Gazette,* 26 June 1885.

16. *PJ* 1, 3 July 1885; *WER,* 1, 2, 3 July 1885.

17. *WER,* 2, 3 July 1885. Garlock, *Guide,* 39, 41, 194, 196, 202, 206, 288–9, 318, 326, 444, 478.

18. *PJ,* 3 July 1885; *WER,* 22, 24 April, 20 June, 1 Sept. 1885; *Framingham Gazette,* 24 April, 15 May 1885. Hoadley spoke at the DA 30 quarterly meeting in Haverhill in July, *Worcester Daily Times (*hereafter cited as *WDT),* 22 July 1885. Richard Oestreicher observed that "as the recognized spokesmen for labor, local and district leaders of the Knights were repeatedly brought in to lead strikes and serve as negotiators, directing great publicity to the order and reaping the organizational credit for the Knights." "Terence Powderly, the Knights of Labor, and Artisanal Republicanism," in Melvyn Dubofsky and Warren Van Tine, eds., *Labor Leaders in America* (Chicago: University of Illinois Press, 1987), 47. Robert Weir captured the overwhelming logistics of an exploding movement: "It is impossible to exagger-

ate the euphoria the KOL's 1885 victory over Jay Gould generated. Coupled with lesser strike and boycott victories in 1885 and early 1886, hundreds of thousands—perhaps more than a million—of workers stormed the KOL veils and begged entry," *Labor's Veil*, 327. Maury Klein, *The Life and Times of Joy Gould* (Baltimore: Johns Hopkins University Press, 1986), 357–63.

19. *PJ*, 3 July 1885; WER, 7, 8, 9 July 1885; *The People*, 29 May 1886.

20. *WER*, 14 July 1885; *PMS*, 15 July 1885.

21. *WER*, 14 July, 21 Aug., 2 Sept. 1885; *PMS*, 15 July, 15 Sept. 1885; *PJ*, 16 July, 3 Oct. 1885. Strikers claimed dozens more left, "driven out by persecution or unwarranted fines." Joe Doherty, *WC*, 1 March 1998.

22. *PMS*, 16, 17 July 1885; *WER*, 20 July 1885. The victory over Jay Gould opened the floodgates to new members. Frederick Turner, the general secretary-treasurer of the Knights, wrote to Terence Powderly that "I am plowed under with letters some of which are trash. Nearly 300 Assemblies organized in December [1885]. Can you wonder at us falling behind with business. I tell you what if this continues the Board will have to be in session continuously." Turner to Powderly, 4 Jan. 1886, *Terence Powderly Papers* (hereafter cited as *TPP*), Catholic University of America, microfilm, University of Massachusetts, Amherst, Reel 12. George E. McNeill, *The Labor Movement: The Problem of Today* (New York: The M.W. Hazen Co., 1888), 423. Weir, *Labor's Veil*, 327. Charles W. Hoadley to *WER*, 18 July 1885. Hoadley opened up a post office box in Woonsocket, *WER*, 11 Aug. 1885. There had been many previous attempts to stabilize costs, rubber imports, and even fix prices among rubber concerns, starting with the loosely drawn Associated Rubber Shoe Companies, 1874 to 1886. Nancy Paine Norton, "The Goodyear Metallic Rubber Shoe Company," Ph. D. thesis (Radcliffe College, 1950), 459; *Framingham Gazette*, 26 June 1885.

23. *PJ*, 27 July, 14 Aug. 1885; *WER*, 23 July, 18 Aug. 1885; *PMS*, 13, 14 Aug. 1885. The Woonsocket Grocers' Association had done some shunning of its own when it issued a customer blacklist in 1882, citing several hundred working-class scofflaws in alphabetical order, *Black List*, n.d., n.p. 1882. There were no other identifying features or exposition in the 57-page pamphlet located at the R.I. Historical Society Library. The *WDT* editorialized that "the strikers at the Woonsocket rubber works still succeed in bulldozing the boarding house and hotel keepers out of feeding the scabs," 15 Aug. 1885. The *PJ* opined that all boycotts were illegal and a single participant in a union was a legal indictment of all members, 16 July 1885. *The Irish World and American Industrial Liberator*, 29 Aug., 26 Sept. 1885. Both the Molly Maguires and the Pennsylvania mine owners had their own boycott lists, Kevin Kenny, *Making Sense of the Molly Maguires* (New York: Oxford University Press, 1998), 105. For a detailed look at boycotts and the Irish, see Michael Gordon, "The Labor Boycott in New York City, 1880–1886," *Labor History* 16 (spring, 1975), 194. New Jersey rubber workers employed a boycott in 1886, Voss, *American Exceptionalism* 127, 170–71;

Scott Molloy, "The Irish in Rhode Island: A Long Struggle to Enter the Mainstream," *P.J.*, 17 March 1997.

24. *PJ*, 23, 24, 27, 31 July, 1, 10, 19 Aug., 7, 15, 18 Sept. 1885; *WER*, 20, 23, 27, 31 July, 16, 23 Sept. 1885; *WDT*, 14, 23 Sept. 1885; PMS, 15 July 1885. Commonwealth of Massachusetts, *Annual Report of the Chief of the District Police for the Year Ending December 31, 1885,* Public Document No. 32 (Boston: Wright and Potter, 1886). Letter from Americus Welch, Chairman, Blackstone Selectmen, to Rufus Wade, Mass. District Police Chief, thanking him for the manpower from Worcester, 30 Nov. 1885, 9–10. The Worcester factory inspector served as a policeman in Millville. Thanks to John Potter for this citation. Kenny, *Molly Maguires,* 281. *Worcester Evening Gazette,* 23 Sept. 1885. The mayor said police were requested by both the company and the union. Hines denied this in a telegram. One of the three Blackstone selectmen was a member of the Knights, *WDT*, 15 Sept. 1885. James Mellen wrote, "We presume Mayor Reed concluded that as two of the Blackstone selectmen were Irish that they must be strikers." *WDT*, 15 Sept. 1885. The hostile *Worcester Sunday Telegram* admitted that the public was still behind the walkout, 20 Sept. 1885. Ironically, Steuart Trench, the agent on the Shirley Estate in County Monaghan and the author of the book on Irish life in that area and era, was often cited against the Mollies during the various trials in Pennsylvania, Kenny, *Molly Maguires,* 14–15. See his chapter 6 for the transformation of the secret society into the Knights of Labor in the eyes of the authorities; also p. 281.

25. *WER*, 4, 5, 10, 17, 26, 27 Aug. 1885; *PJ*, 10, 19 Aug., 7 Sept. 1885. Banigan's boot production never surpassed 10 percent of the usual daily production of 500 pairs during the strike. *PMS*, 30 Sept. 1885. Rumor had it that Banigan received a telegram from a supervisor just before the meeting telling him to hold the line against the strikers because the walkout was ending. Joe Doherty, *WC*, 19 April 1998.

26. *PV*, 30 July 1898; *WER*, 17 Aug. 1885.

27. Joseph Banigan to Terence Powderly, 20 Aug. 1885; *TPP*, Reel 10. Terence Powderly, *The Path I Trod: The Autobiography of Terence V. Powderly* (New York: Columbia University Press, 1940), 104–16; *WER*, 26, 27 Aug. 1885.

28. *PMS*, 19 Sept. 1885; *WER*, 18, 19 Sept. 1885; *PJ*, 11 Dec. 1887. Powderly was an Irish-American activist: Powderly, *The Path I Trod*, ch. 14, "In Ireland's Cause." According to Oestreicher and Philip Foner, Powderly sought the approbation of the elite. Oestreicher, "Terence Powderly" 42; Philip Foner, *History of the Labor Movement in the United States: From the Founding of the American Federation of Labor to the Emergence of American Imperialism* (New York: International Publishers, 1955), 168–69. The general master workman frequently threatened to resign, especially over the issue of illegitimate strikes. Faye Dudden, "Small Town Knights: The Knights of Labor in Homer, New York," *Labor History* 28 (summer, 1987), 315. He was also a stickler for constitutions and bylaws, Samuel Walker. "Terence V. Powderly, Machinist,

1866–1877," *Labor History* 19 (spring, 1978), 177. The Massachusetts Knights of Labor, by and large, endorsed arbitration over strikes, not unlike Powderly himself, Jama Lazerow, "The Workingmen's Hour: The 1886 Labor Uprising in Boston," *Labor History* 21 (spring, 1980), 211. Strangely, there is little mention of ethnicity in this account. In support of Powderly, the use of strikes was not an integral part of the Order's official vision of reforming society. Norman Ware, *The Labor Movement in the United States, 1860–1895: A Study in Democracy* (New York: Appleton and Co., 1929), xvi.

29. The flyer is in *TPP*, Reel 10.

30. *WER*, 29 Aug., 1 Sept., 20 Nov. 1885; Banigan to Carlton, 8 Sept. 1885, partially reprinted in *WER*, 18 Sept. 1885; *WDT*, 30 Sept. 1885; Dudden, "Small Town Knights," 309. Banigan again displayed his knowledge of the Knights' constitution, which contained a general prohibition against membership for liquor dealers. The Massachusetts Commissioner of Labor stated that $2,500 had been collected. *Nineteenth Annual Report, 1888* (Boston: Wright and Potter, 1888), 28–29; *Framingham Gazette*, 28 Aug. 1885; *Haverhill Reporter*, 5 Sept. 1885; *Eighth Annual Report*, District Assembly No. 30, Knights of Labor, Jan. 19–20, 1886 (Boston: Co-operative Printing and Publishing Co., 1886), 38. Thanks to Bruce Cohen for this citation. See Timothy Meagher's index for more information about Mellen.

31. *PJ*, 17 June, 11, 28 Aug., 3, 4, 7, 8, 9, 13, 26 Sept., 17 Oct. 1885; *WER*, 10, 19 Sept. 1885; *WDT*, 16 Sept. 1885. Banigan allegedly told strikers they could stay in the tenements as long as they paid rent. Although the housing units were described as being in varying conditions, some have stood the test of time and remain in use today.

32. *PJ*, 10, 13, Sept. 1885; *WER*, 4, 10 Sept. 1885; *WP*, 3, 10, 22 Nov. 1882; *Uxbridge Compendium*, 18 Sept. 1885. The note to the shoe dealers is in *WDT*, 26 Sept. 1885. Joshua L. Rosenbloom, "Strikebreaking and the Labor Market in the United States, 1881–1894," *Journal of Economic History* 58:1 (March 1998).

33. *WER*, 14 Sept. 1885; *PMS*, 15 Sept. 1885; *PJ*, 15 Sept. 1885. Apparently none of the outside strikebreakers were rubber workers, *WDT*, 23 Sept. 1885. In the rubber strike in Bristol, R.I. the native-born Yankee workers were described as having "class reputations for sobriety and steadiness," *PJ*, 5 Sept. 1885. Litchman served on an Industrial Commission that queried the founder of the United States Rubber Company, Charles R. Flint, in 1901. Flint, when asked about strikes, claimed never to have faced any in the many industries he controlled. *Report of the Industrial Commission*, 57th Congress, 1st Session, Vol. 13, 85. For more detail on Litchman see Weir, "When Friends Fall Out," 103–23. The author identified Foster, McNeill, and Carlton as in favor of trade unions while Litchman stood against the craft mentality.

34. *WP*, 29 Aug. 1879, 24 Dec. 1880; *PJ*, 5 April 1889; *PMS*, 9 July 1878; *WER*, 18 July 1890; James W. Smyth, *History of the Catholic Church in Woonsocket and*

Vicinity, from the Celebration of the First Mass in 1828 to the Present Time, with a Condensed Account of the Early History of the Church in the United States, (Woonsocket: Charles E. Cook, Printer, 1903), ch. 12. McCabe said that John Holt, Banigan's father-in-law, had opposed the wage reduction. A Methodist minister in Uxbridge preached against the evils of striking and endorsed arbitration, *Uxbridge Compendium*, 9 Oct. 1885. Powderly sighed that "I wish to heaven that our members were as well instructed on the labor question as the Catholic clergy," cited in Browne, *Church and Knights*, 83.

35. *PJ*, 7 April 1885; Hayman, *Catholicism in Rhode Island* 1:154, 286–87; *PMS*, 19 June, 9, 12 July 1878.

36. *PMS*, 14 Sept. 1885; *WER*, 15, 17 Sept. 1885.

37. *WER*, 17 Sept. 1885; *WDT*, 22 Sept. 1885. The *PMS* claimed that McCabe was a stockholder in the Woonsocket Rubber Company. He did hold stock in his name but on behalf of the R.I. Catholic Orphan Asylum, *WP*, 26 March 1897. McCabe had written a laudatory letter about Banigan and the Home for the Aged in *WER*, 31 May 1884; *PV*, 26 Sept. 1885. Charles R. Flint to Joseph Banigan, 12 Sept. 1895, Diocese of Providence, *Archives*.

38. *WER*, 17, 18, 21 Sept., 12 Oct. 1885. *PJ* editorial reprinted in *WER*, 18 Sept. 1885; Kenny, *Molly Maguires*, 159; Emmet Larkin, "The Devotional Revolution in Ireland, 1850–75," *American Historical Review* 77 (June 1972), 625–52.

39. *PJ*, 17, 18 Sept. 1885; *WER*, 21 Sept. 1885.

40. *WER*, 18 Sept., 2 Oct. 1885; Doherty, *WC*, 31 May 1998. Carlton wrote to Powderly: "when we really enter upon a battle everything that can be used should be and if force is to be used every conceivable kind of a force at our disposal should be held in readiness for the fray." American Bureau of Industrial Research: *Manuscript Collections in the Early American Labor Movement, 1862–1908. A. A. Carlton Letter Books*, 9 Feb. 1888, Reel 10, Microfilm, University of Massachusetts, Amherst. Rubber workers at the original Woonsocket Rubber Company location wildcatted in 1875 over a wage cut. The bootmakers settled quickly but "the girls" held out longer, protesting reductions and "the very disagreeable smell of the rubber stock on and about them." *WP*, 27 March 1875, 30 Dec. 1880, 18 March 1892, 13 July 1894; *PJ*, 16 Feb., 18 Sept. 1885. The Ladies Assembly formed on July 23, 1885, *Annual Report of the Commissioner of Labor, 1888* (Providence: E.L. Freeman and Son, 1888), 76. The Rhode Island Knights reported 364 female rubber workers in Woonsocket, *The People*, 29 May 1886.

41. *WER*, 2 Oct. 1885; *Journal of United Labor*, 25 Sept. 1885.

42. *WER*, 2 October 1885; *PJ*, 19, 25 Sept. 1885; *The Knight of Labor* (Lynn, Mass.), 10 Oct. 1885. The *PJ* had recently lost an arbitration decision to the Knights who printed the paper, *PMS*, 18 Oct., 19, 25 Nov. 1885; Doherty, *WC*, 24 May 1998.

43. *WER*, 2, 5, 8, Oct. 1885; *PJ*, 4 June 1864, 9 Oct. 1885. A copy of the original is in the McFee Memorial Library, Crowley-Bacon Room, Woonsocket

High School, under *Banigan File*. The Goodyear Metallic Shoe Company in Connecticut required church attendance, Nelson, *American Rubber Workers*, 11. The regulations drew no protest at the Woonsocket facility where workers signed the document, *WER*, 20 Oct. 1885.

44. *WER*, 9 Oct. 1885; *PJ*, 9 Oct. 1885. A Protestant organization in 1902 pegged Catholic Church attendance in Providence at 87 percent, Evelyn S. Sterne, "Whose Church Is It? Conflict over Space in Providence, R.I., 1900–1930," paper presented at the North American Labor History Conference, Wayne State University, 23 Oct. 1997. *WDT*, 9 Oct. 1885; *PMS*, 10 Oct. 1885. See Sydney V. James, *The Colonial Metamorphoses in Rhode Island: A Study of Institutions in Change* (Hanover: University Press of New England, 2000), for the early beginnings of separation of church and state. Mark S. Schantz, *Piety in Providence: Class Dimensions of Religious Experience in Antebellum Rhode Island* (Ithaca: Connell University Press, 2000), 259–60.

45. *WER*, 9, 10, 12 Oct. 1885.

46. *WER*, 16 Oct. 1885; Doherty, *WC*, 7 June 1998.

47. *PJ*, 14, 17, 22, 23 Oct. 1885; *WER*, 12, 17, 20 Oct. 1885.

48. *WP*, 23 Oct. 1885; *WER*, 2 Nov. 1885.

49. *PJ*, 17, 22, 23 Oct. 1885; *WER*, 17, 20 Oct. 1885. A cynic commented that "none but good Christians voted in Rhode Island last Wednesday as all who did vote made the sign of the cross before depositing their ballots." Banigan's children were all around the age of twenty during the strikes, Berenice H. Blessing, "Joseph Banigan: Rhode Island's First Catholic Immigrant Millionaire: A Sketch of a Forgotten Man," (unpubl. undergrad paper, Providence College, 1979), 36–37; William McLoughlin, *Rhode Island: A Bicentennial History* (New York: Norton, 1978), 161–62; *WER*, 1 Sept. 1885; *WP*, 4 April 1890. Woonsocket voted narrowly for a city charter the same year as the constitutional amendment that also allowed such urban incorporations. Robert Bellerose, *Woonsocket* (Dover, N.H., 1997), 11; David Montgomery pinpointed the 1880s as the period when Irish Americans achieved prominence in the unions, "The Irish and the American Labor Movement," in David Doyle, ed., *America and Ireland* (Greenwood, Conn. Greenwood Press, 1980), 206, 214. Rhode Island became the first state with a Catholic majority but remained the last with a partial property qualification for Whites until 1978. Alexander Keyssar, *The Right to Vote: The Contested History of Democracy in the United States* (New York: Basic Books, 2000), 130.

50. *PV*, 30 July 1898.

51. *PV*, 30 July 1898.

CHAPTER 6: TRAGEDY, PHILANTHROPY, AND LACE CURTAIN (pages 140–67)

1. Berenice Blessing, "Joseph Banigan: Rhode Island's First Catholic Immigrant Millionaire: A Sketch of a Forgotten Man," (unpubl. undergrad.

Paper, Providence College, 1979), 36; *PJ*, 23 May 1864; Edward T. Mc-
Carron, "Altered States: Tyrone Migration to Providence Rhode Island
During the Nineteenth Century," *Clogher Record* 16:1 (1997), 156.

2. Blessing, "Banigan," 36.

3. Margaret W. Carter, *Shipwrecks on the Shores of Westerly* (privately printed, ca.
1973), 34–38; the book contains a copy of an original letter from the captain of the Watch Hill Lifesaving Station, which launched a rescue boat that
saved some passengers, Daniel F. Larkin to R. B. Forbes, Watch Hill, 30
Dec. 1872. A letter to the *Woonsocket Patriot* also provided a first-hand account by another passenger, *Woonsocket Patriot* (hereafter *WP*), 6 Sept.
1872; Charles Carroll, *Rhode Island: Three Centuries of a Democracy*, 4 vols.
(New York: Lewis Historical Publishing, 1914), 2:834.

4. *WP*, 6 Sept. 1872; Carter, *Shipwrecks*, 34; Carroll, *Rhode Island*, 2:834; The
Providence Journal (hereafter *PJ*) listed Banigan among the survivors, 2 Sept.
1872.

5. Banigan obituary, *PJ*, 29 July 1898.

6. Stephen Boothroyd, "Richmond Switch," Danger Ahead!: Historic Railway
Disasters; http://danger-ahead.railfan.net/accidents/richmond_switch/
home.html.

7. *WP*, 25 April 1873; Boothroyd; *PJ*, 21 April 1873.

8. Edmund S. Morgan, *Visible Saints: The History of a Puritan Idea* (New York:
New York University, 1963).

9. Robert W. Hayman, *Catholicism in Rhode Island and the Diocese of Providence,
1886–1921* (Providence: Diocese of Providence, 1995), 2: ch. 1; Patrick T.
Conley, "Matthew Harkins: The Bishop of the Poor," in Patrick T. Conley,
Rhode Island in Rhetoric and Reflection: Public Addresses and Essays (East Providence: Rhode Island Publications Society, 2002), 367–87.

10. Hayman, *Catholicism*, vol. 2; Conley, "Harkins: Bishop of the Poor."

11. Patrick T. Conley, "Matthew Harkins: Catholic Bishop and Educator," *Rhode
Island History* 53:3 (1995), 72.

12. Hayman, *Catholicism*, 2:1–5, 726, 730–32, 735; Patrick T. Conley, "Bishop
Matthew Hankins: A Study in Character," in *Rhetoric and Reflection* 364, 366.

13. *Commemorating the One Hundredth Anniversary of the Founding of St. Joseph's
Parish, Hope Street, Providence, Rhode Island, 1851–1951*, n.a. (Providence:
privately printed, 1951), n.p.

14. *Providence Visitor* (hereafter *PV*), 31 May 1884.

15. Hayman, *Catholicism*, 1:261; Blessing, "Banigan," 19; "Diary," Bishop Harkins, 21 Jan. 1898, Diocesan *Records; PV*, 7 June 1884, 27 April 2006; *PJ*, 29
April 2006. The contemporary newspaper accounts reported the 125th anniversary of the facility in 2006.

16. Blessing, "Banigan," 20; *PV*, 31 May 1884.

17. *PV*, 7 June 1884.

18. *PV*, 7 June 1884.

19. Blessing, "Banigan," 20; Carroll, *Rhode Island*, 2:1025; *Providence Evening*

Telegram, 28 July 1898; Flyer, "Home for the Aged Poor," Diocesan *Records,* n.d.

20. See ch. 4, fn 41; Vatican researchers were unable to discover a letter of recommendation for Banigan although there were some bureaucratic documents relating to the nomination process, Vatican Archives (Archivo Segreto Vaticano), *Papers of Pope Leo XIII,* Sec. Brev. 5830, May 1885, 587, 590, 592. My thanks to URI foreign language professor Mario Trubiano for assisting me in the translation.

21. Conley, "Harkins: Bishop of the Poor," 377–78; *PJ,* 9 Feb, 8 April 1867.

22. Hayman, *Catholicism,* 2:475–76; *Prospectus, Charter and Constitution of the Rhode Island Catholic Asylum, Prairie Avenue, Providence, R.I.* (Providence: J. A. and R. A. Reid, Printers, 1889); Conley, "Harkins: Bishop of the Poor," 372–5.

23. *Prospectus,* 6.

24. Hayman, *Catholicism,* 2:475–76; *Prospectus.* John Freeman, a state representative from the Blackstone Valley, lobbied at the legislature in June 1900 for a $3,000 appropriation for the asylum that lost by a single vote, although the amount had already been halved. Freeman blamed the defeat on the negative stance of one of the assembly's most progressive legislators, future Democratic governor Lucius Garvin. Ironically, Republican Party boss Charles Brayton argued in favor of the grant; Freeman to Harkins, 13 June 1900, Diocesan *Archives;* Conley, "A Study in Character," *Rhetoric and Reflection,* 361.

25. Hanley to Banigan, 18 March 1889, Diocesan *Archives.*

26. Banigan to Harkins, 19 March 1889, Diocesan *Archives.*

27. Harkins, "Diary," 2 Sept. 1887, 24 July, 30 Aug., 4 Sept., 20 Nov., 15 Dec. 1888, 6 Jan. 1889, 10 Jan. 1890; Patrick T. Banigan, *Citizenship Papers,* 5 April 1875, U.S. Circuit Court, Providence, Vol. B, 1847–1884; Hayman *Catholicism,* 2:476–77; *Prospectus,* 3.

28. *PJ,* 2 Nov. 1896; *PV,* 31 Oct. 1896, 18 Dec. 1897; Hayman, *Catholicism,* 2: 478–84; *Twentieth Annual Report of the St. Vincent de Paul Infant Asylum* (Providence: Providence Visitor Press, 1912), 5; Blessing, "Banigan," 26–27.

29. *PJ,* 2 Nov. 1896. The actual dedication of the facility occurred on 17 Dec. 1897.

30. Hayman, *Catholicism,* 2:528–37; Blessing, "Banigan," 27, 40; Banigan *Will;* Conley, "Harkins: Bishop of the Poor," 368–69; *WP,* 1 March 1889.

31. Blessing, "Banigan," 22, 25; Conley, "Harkins: Bishop of the Poor," 378–79; "Mrs. Joseph Banigan Dropped Dead," obituary, *Providence Evening Bulletin* (hereafter *PEB*), 17, 18, 21, 23 Jan. 1901.

32. Ann Galligan, "St Maria's Home for Working Girls," unpub. Undergrad. paper, Brown University, 1977, 10.

33. Hayman, *Catholicism,* 2:502–505; Galligan, "St. Maria's Home."

34. Harkins, "Diary," 25 Nov. 1894.

35. Hayman, *Catholicism,* 2:506; Galligan, "St. Maria's Home;" *Boot and Shoe*

Reporter (hereafter *BSR*), 23 Jan. 1895; Blessing, "Banigan" 24; Banigan, *Will.* Conversation with John Parker whose father, James, worked for Banigan in the 1890s. Author interview, John Parker, 10 May 2006, Warwick, R.I.

36. Hayman, *Catholicism,* 2:504–505; Galligan, St. Maria's Home"; "Ledger Book," St. Maria's Home, Diocesan *Archives; Preserving Providence,* 34:3 (winter, 1997); Providence Preservation Society, *Revolving Fund News* (summer, 1998).

37. *PJ,* 16 Jan. 1895.

38. Hayman, *Catholicism,* 2:14–15; *PJ,* 16 Jan. 1895; *PV,* 12 Jan 1895.

39. *Sixth Annual Report of the Rector of the Catholic University of America* (Washington, D.C.: The Church News Publishing Company, 1895), 8.

40. Catholic University *Bulletin* 2 (1896), 97.

41. *Eighth Annual Report of the Rector of the Catholic University of America* (Washington, D.C.: The Church News Publishing Company, 1897), 7, 27; Blessing, "Banigan," 30–31; The Ancient Order of Hibernians also endowed a chair of Catholic History at the school, Hayman, *Catholicism,* 2:635; *PJ,* 16 Jan. 1895.

42. *Sixth Annual Report of the Rector,* 8.

43. "Will of Mrs. Banigan Filed Today," *PEB,* 23 Jan. 1901. She named only relatives.

44. Hayman, *Catholicism,* 2:665–6, 687; Harkins to Andrews, 19 Aug. 1893, Brown University *Archives,* Hay Library. Theresa Ross, "'The Lengthened Shadow of One Man': E. Benjamin Andrews and Brown University, 1889–1898," *Rhode Island History* 65:1, (winter/spring, 2007), 17.

45. "Room 3 Revisited," 2, *Phi Kappa* Fraternity File, Brown University *Archives,* Hay Library.

46. "Room 3," 13; Banigan to Andrews, 16 June 1896, Brown University *Archives,* Hay Library.

47. "Room 3," 6; Scott Molloy, "Behold the Irish Ascendancy 'beyant the sea,'" *PJ,* 17 March 2006.

48. Pictures and short biographies of these students appeared in the fraternity's magazine, *The Temple,* March 1944, Brown University *Archives,* Hay Library; Scott Molloy, *Trolley Wars: Streetcar Workers on the Line* (Washington: Smithsonian Institution Press, 1996), ch. 10.

49. Undated article from *Temple Magazine,* ca. 1927, sketched the life of Higgins; Mohr, 302.

50. Banigan to Andrews, 16 June 1896, Brown University *Archives,* Hay Library; *Treasurer's Report,* Brown University (Providence: The Providence Press, 1897), 15.

51. *BSR,* 6 March 1889; *WP,* 21 June 1889.

52. *BSR,* 6 March 1889; Frank J. Cannon and Harvey J. O'Higgins, *Under the Prophet in Utah* (Boston: The C. M. Clark Publishing Co., 1911), 141; http://www.saints without halos.com/nbks/1898jan.htm; Terry Coleman,

Passage to America: A History of Emigrants from Great Britain and Ireland to America in the Mid-Nineteenth Century (London: Penguin Books, 1972), 245–47.

53. *Deseret News,* 28 Dec. 1895, from an earlier "discourse" by Smith in October; Leonard J. Arrington, *Great Basin Kingdom: An Economic History of the Latter-Day Saints, 1830–1900* (Cambridge: Harvard University Press, 1958), 390.

54. *Deseret Evening News,* 26 Aug. 1893, cited in Arrington, *Great Basin Kingdom,* 390–91.

55. "Minutes," Pioneer Electric Power Company, 26 Nov. 1895, cited in Arrington, *Great Basin Kingdom,* 395.

56. *Deseret News,* 1, 13 Aug., 10 Dec. 1898; Arrington, *Great Basin Kingdom,* 395–97, 399, also see his footnotes for ch. 13. Blessing, "Banigan," 32–33; *WP,* 27 March 1896, 29 July 1898; Thomas W. Bicknell, *The History of the State of Rhode Island and Providence Plantations,* 5 vols. (New York: American Historical Society, 1920) 5:487.

57. Seebert J. Goldowsky, *A Century and a Quarter of Spiritual Leadership: The Story of the Congregation of the Sons of Israel and David (Temple Beth-El)* (Providence: Congregation of the Sons of Israel and David, 1989), 92, 103; *WP,* 29 July 1898.

58. Philip J. Weimerskirch and Philip G. Maddock, *The Irish Literary Renaissance in Providence* (Providence: Providence Public Library, 1996).

59. Weimerskirch and Maddock, *Irish Renaissance,* 3; Patrick T. Conley, *The Irish in Rhode Island: A Historical Appreciation* (Providence: The Rhode Island Heritage Commission and the Rhode Island Publications Society, 1986), 21–22; Garrett D. Byrnes and Charles H. Spilman, *The Providence Journal: 150 Years* (Providence: Providence Journal Company, 1980), 197–200, 215–16; Horace Reynolds, *A Providence Episode in the Irish Literary Renaissance* (Providence: The Study Hill Club, 1929).

60. Evelyn S. Sterne, *Ballots and Bibles: Ethnic Politics and the Catholic Church in Providence* (Ithaca: Cornell University Press, 2004), 69–70; Conley, *Irish in Rhode Island,* 15–18, 30. For an Irish-American schooling in Rhode Island that did not work out, see the fictional, yet historically accurate, Edward McSorley, *Our Own Kind* (New York: Harper and Brothers Publishers, 1946).

61. Carroll, *Rhode Island,* 2:1150–52; *PJ,* 1 Feb. 1888; *Providence Weekly Review,* 6 Aug. 1870; "Diary," Charles Taudvin, 4 Feb. 1889, RIHS, Mss. 9001; Letter from W. D. Merchant to unnamed "Uncle," 10 March 1872, in author's collection.

62. *American Citizen,* 22 Feb. 1896; Author interview with John Parker, 10 May 2006; Patrick T. Conley, "The Persistence of Political Nativism in Rhode Island, 1898–1915: The A.P.A. and Beyond," in Conley, ed., *Rhetoric and Reflection,* 465–72.

63. Reynolds, *A Providence Episode,* 10; Joel Perlmann, *Ethnic Differences: Schooling and Social Structure Among the Irish, Jews and Blacks in an American City, 1880–1935* (Cambridge: Cambridge University Press, 1988), 48.

64. McSorley, *Our Own Kind.* The novel was a Book Club selection.

CHAPTER 7: RUBBER KING AND RUBBER WORKERS
(pages 168–95)

1. *Providence Journal* (hereafter *PJ*), 2 Nov. 1896.

2. *Woonsocket Patriot* (hereafter *WP*), 12, 19 1886.

3. Glenn D. Babcock, *History of the United States Rubber Company: A Case Study in Corporation Management* (Bloomington: Graduate School of Business, Indiana University, 1966), ch. 2.

4. Cited in Babcock, *Rubber Company*, 23.

5. Babcock, *Rubber Company*, ch. 2.

6. Babcock, *Rubber Company*, 24.

7. Babcock, *Rubber Company*, ch.2; *WP*, 2 Feb., 27 July 1883, 25 Feb. 1887.

8. *WP*, 9, 16 Feb., 13 April, 11 May, 20 July 1883, 22 Feb., 16 May 1884, 7 Jan., 25 March, 10 June, 19 Aug. 1887; *Boot and Shoe Reporter* (hereafter *BSR*), 19 June 1889 (some volumes of this industry magazine are in the Lynn, Mass., Public Library and the New York Public Library); William Woodruff, *The Rise of the British Rubber Industry During the Nineteenth Century* (Liverpool: Liverpool University Press, 1958), 46.

9. *Shipton Green against Woonsocket Rubber Company*, Arbitration Hearing, 9 June 1887, *Colt Papers*, Series 14, Box 64, folder 41, Special Coll., URI; *WP*, 7 Oct. 1887.

10. Ibid., 6.

11. Ibid., 97.

12. Ibid., 8.

13. Ibid. 23.

14. Ibid. 23; *WP*, 7 Oct. 1887.

15. Henry C. Pearson, *The Rubber Country of the Amazon* (New York: The India Rubber World, 1911), 38–40, 201; Pearson uncharacteristically provided a verbatim account of Banigan's speech to the Boston Boot and Shoe Club, *India Rubber World* (hereafter *IRW*), 15 July 1892; *WP*, 27 Jan. 1882. *BSR*, 6 March 1895.

16. Samuel Crowther, *John H. Patterson: Pioneer in Industrial Welfare* (New York: Garden City Publishing Company, 1926), 180–87.

17. H. Theobold to Patterson, 25 Jan. 1898, cited in Crowther, *John H. Patterson*, 182–83.

18. H. Theobold to Patterson, 25, 29 July 1898, cited in Crowther, *John H. Patterson*, 184–85, 180–87. Thanks to Derek Milewski for this citation.

19. *Providence Visitor* (hereafter *PV*), 30 July 1898; Scott Molloy, *Trolley Wars: Streetcar Workers on the Line* (Washington: Smithsonian Institution Press, 1996), 53–56; Blessing, *Banigan*, 39.

20. *Despatches* [*sic*], United States Consuls in Para, Record Group 19, 1870–1920, United States *Archives*, Microfilm, University of Mass., Amherst, Reel 6, 6 Sept. 1890.

21. Ibid.

22. Marianne Schmink and Charles H. Wood, *Contested Frontiers in Amazonia*

(New York: Columbia University Press, 1992), 43–44; Woodruff, *Rise of the British Rubber Industry,* 47–48, 66; Barbara Weinstein, *The Amazon Rubber Boom, 1850–1920* (Stanford: Stanford University Press, 1983), 16, 54, 143–46; Major J. Orton Kerby, *An American Consul in Amazonia* (New York: William Edmund Rudge, 1911), 56, 167–88. One account bifurcated control of the industry between Christians and Jews, *WP,* 25 Dec. 1879. *The India Rubber World* sometimes itemized a ship's rubber cargo by importers, including Banigan. See, for example, 15 Sept. 1892.

23. Woodruff, *Rise of the British Rubber Industry,* 48.
24. *WP,* 4 Jan. 1889.
25. *WP,* 29 July 1898.
26. *BSR,* 6 March 1895.
27. United States Rubber Co. vs. Joseph Banigan, *Colt Papers,* URI, Special Collections, box 64, folder 38, 280.
28. *WP,* 29 July 1898; *PV,* 30 July 1898; Richard M. Bayles, ed., *History of Providence County, Rhode Island,* 2 vols. (New York: W.W. Preston, 1891), 1:683; Thomas W. Bicknell, *The History of the State of Rhode Island and Providence Plantations,* 5 vols. 5:486; *BSR,* 6 March 1895; *IRW,* 15 April 1890.
29. *Providence Journal of Commerce,* Aug., 1898, 228.
30. *BSR,* 5 Feb. 1890.
31. *Board of Trade Journal,* Providence, Dec. 1889, 22.
32. Ibid, March 1890, 160.
33. Ibid, March 1890, 160. At the time the Republican state legislature still picked U.S. senators. Regardless of Banigan's business achievements, the general assembly would not have chosen him. The rubber king probably liked to create controversy and agitate the Republican machine.
34. *IRW,* 15 Nov. 1889, 15 March 1890, 15 March 1891; *WP* 6 Dec. 1889, 7 Feb. 1890.
35. *WP,* 28 Dec. 1888, 4 Jan. 1889.
36. *WP,* 4 Jan. 1889.
37. *WP,* 28 Dec. 1888, 4 Jan. 1889.
38. *WP,* 25 Jan. 1889.
39. Ibid.
40. *WP,* 4, 18, 25 Jan., 1, 15 Feb., 1, 8, 22 March 1889; *BSR,* 9 Jan. 1889.
41. *PV,* 21 Oct. 1889.
42. Ibid.
43. Ibid.; *WP,* 18, 25 Oct. 1889.
44. *Board of Trade Journal,* Feb. 1890; *WP,* 21 March, 24 Oct. 1890; *IRW,* 15 April, 15 Oct., 15 Nov. 1890; *BSR,* 29 Oct., 31 Dec. 1890.
45. All quotations are in *BSR,* 29 Oct. 1890.
46. *WP,* 25 Dec. 1885, 8 April 1887; 1, 29 Jan., 26 Feb., 5, 12 March 1886; *Woonsocket Evening Reporter* (hereafter *WER*), 28 Nov. 1885. John O'Keefe, master workman of DA 99, wrote in the Knights' national magazine, the *Journal of United Labor,* that "the native citizen is deprived of the right to vote for

alderman and council unless he owns real estate," 25 Dec. 1885. A similar statement appeared in *John Swinton's Paper,* 14 Feb. 1886. Immediately after the strike the Millville LA claimed that 58 had not been rehired; the local newspaper estimated that most had been recalled. The Knights' notice to pay the bills of 14 is probably the most accurate number, *WER,* 5, 15, 28 Nov. 1885. For more on McGee's career, *WP,* 12 June 1895. *Providence Morning Star* (hereafter *PMS*) 4 April 1886. *Eighth Annual Report,* District Assembly, No. 30, Knights of Labor, Jan. 19–20, 1886 (Boston: Co-operative Printing and Publishing Co., 1886), 40.

47. *The People,* 29 May 1886; *WP,* 12, 19 March, 11 June 1886, 15 July 1887; *WP,* 6 July 1887, 18 Jan., 16 August 1889. Long after the fact, Joseph McGee would attack town subsidies to the firm, *WP,* 26 March 1897.

48. *WP,* 9 July, 20 Aug., 3, 24 Dec. 1886, 1 Jan., 2 Dec. 1887. *Quarterly Report,* District Assembly, No. 30, Knights of Labor, April 17–20, 1886 (Boston: Co-operative Printing and Publishing Co., 1886), 34. The graduating class of Blackstone High School adopted the union motto *Labor Omnia Vincit.* The Millville chapter of the Ancient Order of Hibernians visited an amusement park in Warwick, Rhode Island, and "depopulated the town." *Annual Report of the Commissioner of Industrial Statistics, 1888* (Providence: E. L. Freeman and Son, 1888), 60–61.

49. *The Haverhill Laborer,* 13 Feb. 1886 (the microfilmed copies are incomplete from the collection of the Wisconsin Historical Society). The *New Haven Union,* 2, 8 March 1886. (I am indebted to Debbie Elkin for tracking down the Connecticut citations.) The Knights of Labor published a weekly newspaper, *The Agitator,* in Naugatuck, Conn., during this time but there are no extant copies. *PJ,* 4 March 1886; *IRW,* 15 Jan. 1891. Blaine's report to the Massachusetts AFL was reprinted in the industry's leading trade journal. It is an encapsulated history of rubber organization in the Knights and the American Federation of Labor. *Quarterly Report,* District Assembly No. 30, Knights of Labor, July 19–22, 1886 (Boston: Co-operative Printing and Publishing Co., 1886), 69; *Quarterly Report* of DA 99 K. of L., 20 July 1886 (Boston: Co-operative Printing and Publishing Co., 1886), 21; *Quarterly Report* of District Assembly 99 K. of L., 30 Oct. 1886 (Boston: Co-operative Printing and Publishing Co., 1886). George E. McNeill, *The Labor Movement: The Problem of To-Day* (New York: The M.W. Hazen Co. 1888), 26. Carleton sent the negative letter, 31. A bootmaker wrote to *The People* with suggestions for the Connecticut conference, calling for an end to instructing scabs: "I spent money learning the bootmaking trade, but I cannot get work because I am locked out of Mr. Banigan's, while men are hired by my side who never saw a rubber factory before," 27 Feb. 1886. Gerald N. Grob, *Workers and Utopia: A Study of Ideological Conflict in the American Labor Movement, 1865–1900* (Evanston, University of Illinois, 1961), 122–24. Marc Stern, *The Pottery Industry of Trenton: A Skilled Trade in Transition, 1850–1929* (New Brunswick: Rutgers University Press, 1994). These Knights secured a national charter.

50. The *Boston Knight,* 19 June, 10 July, 30 Oct. 1886. The scattered issues of this paper, published for about one year, are on microfilm at the Boston Public Library. Norman Ware, *The Labor Movement in the United States, 1860–1895: A Study in Democracy* (Chicago: Appleton and Co., 1929) 160, 180. Ware erroneously stated that the demand for such charters was waning by 1884; to the contrary, 179–80. Grob, *Workers and Utopia,* 121

51. *Records of the 10th General Assembly,* Knights of Labor, Richmond, Va., 1886; *TPP,* Reel 67, 207, 265–66.

52. Grob, *Workers and Utopia,* ch. 6. The Richmond convention came immediately after the "treaty" with the AFL and probably had little influence on the earlier resolutions by the Knights.

53. All the correspondence and commentary are in the *Proceedings of the General Assembly,* Knights of Labor, Eleventh Regular Session, Minneapolis, Minn., Oct. 9–19, 1887; *Powderly Papers*, Reel 67, 1299–1300. Philip Foner, *History of the Labor Movement in the United States,* 2:159–60. WP, 15 April 1887. Jama Lazerow, "The Working Men's Hour: The 1886 Labor Uprising in Boston," *Labor History* 21 (1980), 205–208. Grob, *Workers and Utopia,* 106. *Labor Leader,* 21 May 1887. Dennis J. Starr, "The Limits of Conservative Unionism: The Rubber Strike of 1904 in Trenton, New Jersey," *New Jersey History* 5 (fall-winter, 1988). The author mentioned that workers there had been de-skilled and claimed the walkout was "the first major rubber strike in the United States." He was apparently unaware of Millville and Woonsocket in 1885, 1, 3, 5.

54. *WP,* 2, 16 Sept. 1887, 31 Aug., 7 Sept. 1888,

55. *WP,* 16 Sept. 1887, 31 Aug., 7 Sept. 1888, 14, 21, 28 March, 25 April 1890, 30 Jan. 1891, 18 March 1892, 12 Jan., 11 May, 13 July 1894, 22 May 1896; *PJ,* 11 July 1894. A contemporary Millville historian wrote in 1888 that the strike's economic ramifications "are not yet entirely overcome." Dr. Adrian Scott, "History of Blackstone, Massachusetts up to 1888," in D. Hamilton Hurd, ed., *History of Worcester County Massachusetts* (Philadelphia: J.W. Lewis and Co., 1889). The Millville Knights also held a Christmas dance later in the year, *Labor Leader,* 30 April 1887; Buhle, "The Knights," 58–59.

56. *WP,* 26 Jan. 1894.

57. *IRW,* 15 Jan. 1891; *Labor Leader,* 15 Feb., 11 Oct. 1890, 3 Jan. 1891, 6 Aug. 1892.

58. *WP,* 27 Jan., 10 Feb., 22, 23 March 1888, 17 April, 30 Oct. 1891, 14 Oct. 1892. Buhle, 60–67.

59. *WP,* 15, 22 Dec. 1893, 6 Sept. 1895. The funeral oration is reprinted in James W. Smyth, *History of the Catholic Church in Woonsocket and Vicinity, from the Celebration of the First Mass in 1828 to the Present Time* (Woonsocket: Charles E. Cook, Printer, 1903), 69–81; *PV,* 22 Feb. 1896. "Diary," Bishop Harkins, 14, 18 Dec. 1893, Diocese of Providence, *Archives.*

60. *PV,* 22 Feb. 1896.

61. *WP,* 21 Feb. 1896.

62. *PEB,* 13, 14 March 1896; *PJ,* 17 Feb. 1896; Bayles, *Rhode Island,* 5:400–401.

CHAPTER 8: THE UNITED STATES RUBBER COMPANY
(pages 196–225)

1. V. I. Lenin, *Imperialism: The Highest Stage of Capitalism* (New York, International Publishers, 1939); Nathaniel Wright Stephenson, *Nelson W. Aldrich: A Leader in American Politics* (Port Washington, N.Y.: Kennikat Press, 1971).

2. *Boot and Shoe Reporter* (hereafter *BSR*), 29 Jan. 1890, mentioned Banigan's participation in a New York meeting to form the predecessor to the U.S. Rubber Company, The Rubber Boot and Shoe Manufacturer's Association; *India Rubber World* (hereafter *IRW*), 15 March 1890.

3. *BSR*, 2 Oct. 1889.

4. *IRW*, 15 July 1890; *BSR*, 19 June, 14 Aug. 1889, 15 April, 27 May, 29 July, 26 Aug., 14, 21 Oct., 16 Dec. 1891; Glenn D. Babcock, *History of the United States Rubber Company: A Case Study in Corporation Management* (Bloomington: Graduate School of Business, Indiana University 1966), 24–26; *IRW*, 15 April 1890; "President's Report," United States Rubber Company (hereafter USRCO), 25 May 1896, *Colt Papers*, URI, Series 14, Box No. 60, Folder No. 6.

5. Quotation in *Flint Coll.*, Box 3; Babcock 26–27; Flint authored a fascinating autobiography, *Memories of an Active Life: Men, and Ships, and Sealing Wax* (New York: G. P. Putnam's Sons, 1923); *IRW*, 15 Oct., 15 Dec. 1893. Rumor had it that Banigan also assisted the Brazilian government while he presided over the cartel to squelch a revolution in Para, *IRW*, 10 Feb. 1897; *Woonsocket Patriot* (hereafter *WP*), 21 April 1893. For an incisive Brazilian-Portuguese perspective, consult the interview with Baron de Gondoriz in *IRW*, 15 June 1892. *Nelson W. Aldrich Papers*, Library of Congress, Microfilm available at Providence College *Archives:* Charles Flint to W. E. Curtis, 21 Aug. 1890, Container 27–28, Reel 20; Flint to Nelson W. Aldrich, 10 June 1892, Containers 29–30, Reel 21; Flint to Aldrich, Containers 29–30, Reel 21; Flint to Aldrich, Containers 29–30, Reel 21.

6. Cited in Babcock, *Rubber Company*, 28.

7. Louis H. Bristol to Charles Johnson, 25 June 1892, *USRCO Coll.*, Letters File, Baker Library, Harvard University.

8. Rhode Island Supreme Court, Common Pleas Division, *Joseph Banigan vs. United States Rubber Company*, JTW 319, 30 in *Colt Papers*, series 14, box 64, folder, 38, University of Rhode Island, Special Collections, (hereafter BVUSRCO), 76; Babcock, *Rubber Company*, 27–32; Flint, *Autobiography*, 299–300; *BSR*, 9 March, 6 April 1892; *IRW*, 15 June 1891, 15 May 1894. The incorporation papers are reproduced in *IRW*, 15 April 1892. Flint told of an unnamed bank director who gave him the idea in 1888, Flint to [indistinguishable], 17 March 1888, *Flint Coll.*, Box 1, Folder 1. The collection is small and mostly fragmentary; the *Colt Papers* at URI contain more material.

9. *Report and Proceedings of the Joint Committee of the Senate and Assembly Appointed to Investigate Trusts* (Albany: New York Legislature, 1897), 456.

10. Ibid., 459.

11. Ibid., 501.
12. Ibid., 458–59, 453, 455–56, 470, 501; *Report of the United States Industrial Commission,* (Washington: GPO, 1902), 19 vols., 19:33.
13. Ibid., 19:33, 19:80, 19:82–85, 19:89; Babcock, *Rubber Company,* 27–32; *WP,* 19 Feb. 1897.
14. Babcock, *Rubber Company,* 32–4; *U. S. Rubber Company Coll.,* Folder 1, Baker Library, Harvard University; *BSR,* 2 Nov. 1892; *IRW,* 15 Sept., 1894.
15. Flint, *Autobiography,* 298.
16. *BSR,* 29 June 1892; Babcock, *Rubber Company,* 26, 34–37; Most of Flint's talk is in *IRW,* 15 May 1892.
17. *BSR,* 15 Oct. 1890, 11 May, 7 Sept., 19 Oct., 23 Nov. 1892; *IRW,* 15 April, 15 May 1892; *WP* 24 July 1891.
18. *BSR,* 19 July 1893.
19. *Providence Journal of Commerce,* April 1894, 44.
20. *BSR,* 11 Oct. 1893; *IRW,* 15 May 1893, 15 Feb. 1894; *WP,* 9 Sept., 2 Dec. 1892, 15 Dec. 1893, 4 Sept. 1896. William Banigan received a new house from his father as a wedding gift.
21. *IRW,* 1 Jan. 1899.
22. Henry J. Browne, *The Catholic Church and the Knights of Labor* (Washington, D.C.: The Catholic University of America, 1949), 65.
23. Cited in *IRW,* 15 April 1892.
24. *BSR,* 20 May 1891.
25. New York Legislature, *Report to Investigate Trusts,* 774; Babcock, *Rubber Company,* 31; *BSR,* 1 April 1891.
26. "Purchase Agreement," by the United States Rubber Company for the Woonsocket Rubber Company, 22 March 1893, 4, *U.S. Rubber Co. Coll.,* Baker Library, Harvard University.
27. Ibid., 4.
28. "Purchase Agreement," U.S. Rubber Co., 22 March 1893, Report of Special Committee, Appraisal of Banigan Property, n.d. This investigation, completed with the cooperation of Banigan, was in greater detail than the initial secret estimate, *U.S. Rubber Company Coll.,* Baker Library, Harvard University. For accounts of the wringing business, which also garnered a ten-year tax exemption, see *WP,* 20 Feb., 27 March, 8 May, 18 Dec. 1891. One Boston newspaper reported that Banigan initiated a plan for the absorption of his enterprise, but that may have been a planted item, as Banigan claimed the combination contacted him, *BSR,* 22 March, 26 April, 10 May 1893; *IRW,* 1 June 1909; *WP,* 16 Jan. 1891; *PJ,* 19 April 1893; *BVUSRCO,* 2, 122.
29. Babcock, *Rubber Company,* 36; *IRW,* 15 Feb. 1892.
30. The quotations are in "President's Report," United States Rubber Company, New Brunswick, N.J., 17 April 1894, *Colt Papers,* URI, Series 14, Box 60, Folder 6. The same report exists in the *U.S. Rubber Co. Coll.,* Baker Library, Harvard University, but it was used in a different way. The document

was notarized in 1898 as evidence in a lawsuit between Banigan and the cartel. These initial accounts were mimeographed, *BSR*, 19 June 1895; *WP*, 4 March 1898; *BVUSRCO*, 142, 186, 193.

31. *BVUSRCO*, 273.

32. The quotations are in the "President's Report, 1894"; *WP*, 2 March 1894, 26 July, 6 Sept. 1895; *BSR*, 28 Feb., 7 March, 5 Sept. 1894, 2 Jan. 1895; *BVUSRCO*, 41, 54, 272.

33. The story and quotations are in *IRW*, 15 Aug. 1894.

34. Quotations in "President's Report, 1894"; *WP*, 7 Sept., 4 March 1898.

35. Quotations in the "President's Report, 1894"; *BSR*, 23 Jan. 1895. The latter journal reported erroneously that there was "no lack of harmony," *BSR*, 5 June 1895; *WP*, 3 March 1896.

36. "President's Report, 1894, 1895"; *BSR*, 27 June 1894, 9 Jan., 13 Feb., 29 May 1895.

37. "President's Report, 1895."

38. "President's Report, 1895"; *BVUSRCO*, 134, 136.

39. (Robert Evans), "President's Report, 1896"; *WP*, 11 Jan. 1895, 3 Jan. 1896, 4 March 1898, *BSR*, 12 Feb., 11 March 1896; *BVUSRCO*, 4.

40. *WP*, 17, 31 Jan., 14, 21 Feb., 20, 27 March 1896.

41. *BVUSRCO*, 167–68.

42. *WP*, 20 Dec. 1895, 3 July 1896, 12 Feb., 26 March 1897; *BSR*, 16 Dec. 1896, 20 Jan. 1897, 16 Feb. 1898; Elmer E. Cornwell Jr., "A Note on Providence in the Age of Bryant," *Rhode Island History* 19:2 (1960), 33–40.

43. *BVUSRCO*, 129; Rhode Island Supreme Court, Appellate Division, *Joseph Banigan vs. United States Rubber Company, Brief for the Defendant, William G. Roelker and Samuel Norris Jr., Defendant's Attorneys,* John Hay Library, Brown University.

44. Ibid., 19.

45. Ibid., 19.

46. Ibid., 20–23.

47. Ibid., 212.

48. Ibid., 281.

49. *BVUSRCO*, 44–45.

50. Ibid., 61.

51. Ibid., 96.

52. Ibid., 101–105, 109–21.

53. Ibid., 122.

54. Ibid., 176.

55. Ibid., 176–80.

56. Ibid., 256.

57. Ibid. 255–60.

58. Ibid., 263–70, 283.

59. *BVUSRCO*, 131.

60. *WP*, 14 Feb. 1896.

61. *WP,* 17 Jan., 14 Feb., 15 May, 21 Aug. 1896, 12 March 1897.

62. *WP,* 17 Jan. 1896.

63. *WP,* 12 March 1897.

64. *BSR,* 16 Dec. 1896.

65. *WP,* 10 Aug. 1894, 12 May, 10 July, 21 Aug., 11 Sept., 2 Oct. 1896, 8 Jan., 5, 19 Feb., 12 March, 16 April, 18 June 1897, 7 Jan., 18 Feb., 29 April 1898; *BSR,* 9 Dec. 1896; *IRW,* 1 June 1909; Babcock, *Rubber Company,* 39; "Articles of Association and By-Laws, The Joseph Banigan Rubber Company, Providence, R.I.," *Colt Papers,* Series 14, Box 71, Folder 93.

66. *BSR,* 10 March, 14 April 1897; Babcock, *History of U.S. Rubber,* 39–40; *WP,* 4 Sept. 1896; *BVUSRCO,* 239, 244, 245.

67. *BSR,* 17 March 1897, *IRW,* 1 Feb. 1900.

68. *WP,* 19 March 1897.

69. *WP,* 19 March 1897, 4 Sept. 1896, 16 July, 17 Sept. 1897; Babcock, *Rubber Company,* 39–40; *BSR,* 7 Feb. 1897.

70. "President's Report, 1897."

71. *BSR,* 2 June 1897.

72. *WP,* 4 June, 6 Aug., 3 Dec. 1897, 28 Jan., 6 May 1898; Babcock, *Rubber Company,* 435; *IRW,* 15 July 1892; *BSR,* 11 Aug., 29 Sept., 27 Oct., 3 Nov. 1897, 6 July 1898.

73. Edward C. Stiness, Reporter, *Reports of Cases Argued and Determined in the Appellate Division of the Supreme Court of Rhode Island* (Providence: E. L. Freeman & Sons, 1901), vol. 22, 452–54.

74. *PV,* 30 July 1898.

CHAPTER 9: CONCLUSION: A MEMORIAL FOREVER (pages 226–51)

1. *Boot and Shoe Reporter* (hereafter *BSR*), 21 Oct. 1896, 11 May 1898.

2. *Woonsocket Reporter* (hereafter *WR*), 15 July 1898; *Woonsocket Patriot* (hereafter *WP*), 14 June 1895, 29 July 1898; *Downtown Providence* (Providence: R.I. Historical Preservation Report, 1970), 19, 21–22, 61, 63, 65–66. The skyscraper was known as the Banigan, Grosvenor, and Amica Buildings respectively. Thanks to Roisin Lafferty, who lives in the Dublin area and conducts research about Co. Monaghan, who made a copy of the postcard for me; *BSR,* 3 March 1897; *Providence Journal* (hereafter *PJ*), 24 Jan., 14 April 2002; *Providence Visitor* (hereafter *PV*), 30 July 1898.

3. *BSR,* 2 June 1897; *WR,* 4 June 1897.

4. *Providence Journal of Commerce,* Aug., 1896, Sept., 1898; *WP,* 12 Feb. 1897; *BSR,* 9 Dec. 1896, 21 April, 3 Nov. 1897; *WR,* 3, 17 Sept., 17 Dec. 1897, 19 March 1899; Glenn D. Babcock, *History of the United States Rubber Company: A Case Study in Corporation Management* (Bloomington: Graduate School of Business, Indiana University, 1966), ch. 3, 426–27.

5. The tablet is located at the *Archives,* Sisters of Mercy of the Americas, Cumberland, R.I.

6. Banigan to Sister Mechtilde, Mother Superior, 10 Dec. 1896, *Archives,* Sisters

of Mercy of the Americas. *Seventy-Five Years in the Passing with the Sisters of Mercy, Providence, Rhode Island, 1851–1926,* n.a. (Providence: Providence Visitor Press, 1926), 131.

7. *PV,* 30 July 1898.

8. *PJ,* 29 July 1898.

9. Evelyn S. Sterne, *Ballots and Bibles: Ethnic Politics and the Catholic Church in Providence* (Ithaca: Cornell University Press, 2004), 113–31.

10. *WP,* 21 Jan. 1887, 22 Nov. 1889, 27 Feb. 1891, 1 Jan., 13 May 1892, 7, 14 July, 3 Nov. 1893, 6 April, 19 Oct. 1894.

11. *WP,* 14 Sept. 1894.

12. *WP,* 14 Sept., 5 Oct. 1894, 27 Nov, 1896.

13. *WP,* 18 March 1887, 23 March 1888.

14. *WP,* 22 Feb. 1889.

15. *WP,* 28 March 1890, 20 March 1891, 3 March 1899.

16. *WP,* 18 Nov. 1887, 20 Jan., 7, 14, 27 Dec. 1888, 15 March 1889, 22 May, 30 Oct., 27 Nov. 1891, 21 March, 6 May 1892, 20 Oct. 1893, 25 Jan., 8 March 1895, 6 May 1898, 13 Jan. 1899.

17. *WP,* 10 Feb. 1888, 11 Oct. 1889, 20 March 1891. Peter F. Stevens, *The Voyage of the Catalpa: A Perilous Journey and Six Irish Rebels' Escape to Freedom* (New York: Carroll and Graf Publishers, 2002).

18. *PJ,* 1 Feb. 1888; *WP,* 9, 16 March, 6 July, 3 Aug., 19 Oct. 1888, 8 Feb., 24 May 1889, 18 April 1890. A multicultural class at Central Falls High School in Rhode Island, under the direction of George McLaughlin, undertook a study of Wilson's exploits and made several presentations and a mimeographed pamphlet, "The Search for James Wilson," in 2002. An article about the students' efforts appeared in the *Pawtucket Times,* 22 Nov. 2002.

19. *WP,* 9 Aug., 6, 27 Sept. 1889, 21 Aug., 18 Sept. 1891, 28 July 1893, 2 Nov. 1894, 13 May 1898, 1 Sept. 1899; Patrick T. Conley, *The Irish in Rhode Island: A Historical Appreciation* (Providence: The Rhode Island Heritage Commission and The Rhode Island Publications Society, 1986), 22.

20. James W. Smyth, *History of the Catholic Church in Woonsocket and Vicinity, from the Celebration of the First Mass in 1828, to the Present Time* (Woonsocket: Charles E. Cook, Printer, 1903), 147–58; *WP,* 4 Feb., 4 Nov. 1898, 27 Jan., 14 July, 11, 13, 18 Aug., 10 Oct., 5, 26 Jan., 23 Feb., 9 March 1900; *WR,* 10 Sept. 1897, 13 Jan. 1899, 18 Aug. 1900.

21. *WP,* 5 Aug. 1898.

22. *WP,* 5 Aug. 1898.

23. *PV,* 8 Aug. 1898.

24. *PV,* 8 Aug. 1898; *WP,* 5 Aug. 1898.

25. *PV,* 8 Aug. 1898; *WP,* 5 Aug. 1898; Thomas w. Bicknell, *The History of the State of Rhode Island and Providence Plantations,* 5 vols. (New York: American Historical Society, 1920), 5:488.

26. *PJ,* 29 July 1898.

27. *Providence Morning Star* (hereafter *PMS*), 26 June 1877.

28. Robert W. Hayman, *Catholicism in Rhode Island and the Diocese of Providence* (Providence: Diocese of Providence, 1995), 2:616, 620; John S. Gilkinson, Jr., *Middle-Class Providence, 1820–1940* (Princeton: Princeton University Press, 1986), 144–5.

29. *PV,* 30 July 1898.

30. Bicknell, *Rhode Island,* 5:488.

31. *PJ,* 7 Jan. 2006.

32. *Providence Journal of Commerce,* Sept. 1898.

33. *WR,* 14 April 1899, 18 Aug. 1900; *BSR,* 10 Aug. 1898; Bicknell, *Rhode Island,* 5:487.

34. "Minority populations grow," *PJ,* 26 Dec. 1997; "Which State's the Most Italian?" *PJ,* 9 Aug. 2001.

35. "Minority"; "Which State's"; Conley, *Irish in Rhode Island;* Sterne, *Ballots and Bibles,* ch. 4; Ian S. Haberman, "The Rhode Island Business Elite, 1895–1905: A Collective Portrait," *Rhode Island History,* 26 (1967), 38.

36. Kerby Miller, *Emigrants and Exiles: Ireland and the Irish Exodus to North America* (New York: Oxford University Press, 1985), 266–71; Charles Carroll, *Rhode Island: Three Centuries of a Democracy* (New York: Lewis Historical Publishing, 1932), 2:1149–50.

37. *PV,* 30 July 1898; *PJ,* 29 July 1898.

38. Bicknell, *Rhode Island,* 5:488; Henry Mann, *Our Police: A History of the Providence Force from the First Watchman to the Latest Appointee* (Providence: J.M. Beers, 1889), 230.

39. *PV,* 30 July 1898.

40. *New York Times,* 22 July 1893.

41. *WR,* 29 July 1898.

42. *PV,* 30 July 1898.

43. *WR,* 29 July 1898.

44. Author interview, John Parker, 10 May 2006, Warwick, R.I.

45. Author interview, John Parker, 8 June 2006, Warwick, R.I.

46. *WP,* 29 July 1898.

47. *WR,* 11 May 1894.

48. *WR,* 29 July 1898.

49. Author interview, John Parker, 10 May 2006, Warwick, R.I.

50. Author interview, John Parker, Warwick, R.I., 8 June 2006. Samuel Chapman Smart, *The Outlet Story, 1894–1984* (Providence: Outlet Communications, Inc., 1984).

51. *India Rubber World* (hereafter *IRW*), 1 Aug. 1898.

52. *PV,* 30 July 1898.

53. Ibid.

54. *BSR,* 6 March 1895.

55. Bicknell, *Rhode Island,* 5:485.

56. *PV,* 30 July 1898.

57. Ibid.

58. Bicknell, *Rhode Island*, 5:488.
59. *Providence Evening Bulletin* (hereafter *PEB*), 17, 18 Jan. 1901; *PV,* 19 Jan. 1901.
60. *PV,* 26 Jan. 1901; *WP,* 5 Aug. 1898.
61. *PV,* 2 March 1901.
62. *PJ,* 1 Jan. 1909; Bicknell, *Rhode Island,* 5:488; Berenice H. Blessing, "Joseph Banigan: Rhode Island's First Catholic Immigrant Millionaire: A Sketch of a Forgotten Man" (unpub. undergrad. paper, Providence College, 1979), 37.
63. Blessing, "Banigan," 36–37; *PJ,* 20 June 1900, featured the stained glass windows.
64. Telephone interview with John Banigan, 7 May 2006; Blessing, "Banigan," 20–21.
65. *WP,* 26 Feb. 1892, *PJ,* 9 Sept. 2006; *PV,* Aug. 1972.
66. *PJ,* 17 Feb 1894, 5 July 1899; *Minute Books,* Sarsfield Literary Association, Pawtucket, R.I., 29 Oct., 21 Nov. 1893, 17 Feb. 1894, 28 May 1899, *Special Collections,* URI Library. Albert McAloon, an Irish-American activist, recently donated to the University of Rhode Island the record books of a number of late-nineteenth-century local Hibernian groups in the Pawtucket area.
67. *Providence Evening Telegram* (hereafter *PET*), 28 July 1898.
68. *PJ,* 17 Feb. 1894, 5 July 1899; Banigan was inducted into the Rhode Island Heritage Hall of Fame in 2005 by the author.

INDEX

Candee Rubber Company, 173, 199, 201, 211, 215
Cannon, Frank, 160, 161
Cannon, George Q., 160, 161
Carlton, Albert A., 115, 125, 127, 133, 134, 188
Carpenter, Thomas, 52, 57, 58
Carrickmacross, Ireland, 20
Carrington, Edward, 39
Carroll, Charles, 44, 62, 76, 84, 165, 242
Catalpa (ship), 233
Cathedral of Sts. Peter and Paul, 58, 156, 182
Catholic Emancipation. *See* O'Connell, Daniel
Catholic Knights of America, 204, 233
Catholic Ladies Reading Society, 232
Catholic Orphan Asylum, 149
Catholic University, 156–57, 165, 235
Cavendish, Lord Frederick, 251
Central Labor Union (Rhode Island), 189
Chaffee, Edwin, 69, 70, 72
Charles I, 31
Charles II, 31
Charles Parnell Defense Fund, 130
child labor, 33, 36, 43, 60–62, 95, 104
Childs, Horace, 120
China, 91, 180
Chinese, 76
Church of England, 12, 13, 17, 30, 160
Civil War, 73–76, 130
Claflin & Company, 62
Clan na Gael, 103, 120, 233
Clarke, John, 31
Clifford Street, Providence, 68
Clinton Street, Woonsocket, 81
Clogher, Ireland, 65
coffin ships, 27

Colchester, Connecticut, 125, 172
College Hill, Providence, 158, 177
Colorado Catholic, 162
Colt, Samuel, 98, 187, 198, 201, 204–8, 212, 215–18, 222, 229, 240
Columbus, Christopher, 68
Commercial National Bank, 236
Congregation of Propaganda, 102
Conley, Patrick J., 185
Conley, Patrick T., 34, 146, 164
Connecticut, 30, 46, 117, 143, 182, 189
Connemara, Ireland, 25
Converse, Elisha S., 173, 198, 202, 222, 236
Conway, John, 180
Cook, Lyman, 81, 84, 91, 95, 96, 97, 185
Cook, Simeon, 81
Corcoran, John, 75
Corcoran, Thomas G., 159
Corcoran, Thomas P., 159
Cork, Ireland, 78
Corky Hill, Providence, 35
Corry, Rev. John, 48
cottier, 17, 22
County Carlow, 39
County Cavan, 26
County Donegal, 7
County Kilkenny, 35, 37, 126
County Leitram, 127
County Limerick, 37
County Monaghan, 11–28, 35, 40, 41, 50, 61, 64, 65, 80, 101, 117, 137, 144, 182, 221, 227, 242, 254n
County Tipperary, 51
Cranston, Rhode Island, 50, 51, 78, 150
Cromwell, Oliver, 31
Crowther, Samuel, 175
Crusaders of the Holy Cross, 233
Cumberland, Rhode Island, 85

New York, Providence, and Boston
Railroad, 143
New York Times, 66, 244

O'Connell, Daniel, 17
O'Farrell, Bridget, 37
O'Farrell, John, 37
Olneyville, Rhode Island, 90, 156,
212, 240
Orangemen, 85
Orange Street, Providence, 64, 68,
227
Orchard Street, Providence, 238
O'Reilly, Bishop Bernard, 65, 67
O'Reilly, Rev. Bernard, 7
Our Lady of Providence Cemetery, 249
Our Own Kind, 167
Outlet Company, 247

Panic of 1837, 43
Panic of 1873, 89, 90, 104, 105
Panic of 1893, 194, 210, 226
Para, Brazil, 68, 173, 177, 179, 198
Para Rubber Company, 115, 116, 118,
123, 125, 194, 206
Paris Commune, 120
Parker, James M., 246
Parnell, Charles Stewart, 251
Patterson, John H., 175–77
Pawtucket, Rhode Island, 74, 85,
131, 141, 147
Pearson, Henry C., 175
Peel, Sir Robert, 24
Penal Laws, 13
Pennsylvania, 99, 103, 120, 132
People's Constitution, 47, 49
Perkins, Frances M., 97
petroleum gas, 91
Phi Kappa Sigma, 158, 159
Philadelphia, Pennsylvania, 10, 31, 99
piecework, 110
Pinkerton, Alan, 120
Pioneer Electric Power Company,
160, 161

Plymouth, Massachusetts, 30
Poland, 119
Pond, Mayor Daniel, 185
Poor Union, 24
Pope Gregory XVI, 149
Pope Leo XIII, 12, 102, 145, 148–49
Portsmouth, Rhode Island, 36
Portugal, 178
Potato Famine, 3, 11, 13–26, 35, 50,
59, 86, 117, 137, 140, 141, 149,
160, 162, 165, 206, 225, 230, 237
Potter, Jr., Elizah R., 48
Powderly, Terence V., 99, 113,
121–23
pre-Famine Irish, 21, 34–43, 56
Priestly, Joseph, 68
primary source material, 4–5, 8,
253n, 256n
Progressive Era, 205, 213
Protestant ascendancy, 12, 17
Protestant Social Gospel, 177
Providence and Worcester Railroad,
85, 125, 183
Providence Building Company, 227,
236
Providence Cable Tramway, 177, 236
Providence Chamber of Commerce,
203, 238
Providence College, 146
Providence Commercial Club, 202
Providence Evening Telegram, 203, 236
Providence Fifth Ward. *See* Irish Fifth
Ward, Providence
Providence Journal, 1, 16, 34, 48, 66,
76, 90, 105, 118, 119, 130, 131,
132, 134, 135, 163, 165, 236
Providence Lying-In Hospital, 153
Providence Morning Star, 122, 135
Providence Police Department, 55
Providence Preservation Society, 154
Providence River, 64
Providence Rubber Company, 68, 69,
71, 96, 141
Providence shoes, 69